Joseph Jackson

Flora of Worcester County

A catalogue of the phaenogamous and vascular cryptogamous plants of

Worcester County, Masachusetts. Second Edition

Joseph Jackson

Flora of Worcester County
A catalogue of the phaenogamous and vascular cryptogamous plants of Worcester County, Masachusetts. Second Edition

ISBN/EAN: 9783337268527

Printed in Europe, USA, Canada, Australia, Japan

Cover: Foto ©berggeist007 / pixelio.de

More available books at **www.hansebooks.com**

Rhododendron *Rhodora*. Don.

The *Rhodora*.

*"In May when sea-winds pierced our solitudes
I found the fresh Rhodora in the woods."*

—R. W. EMERSON.

FLORA OF WORCESTER COUNTY.

A CATALOGUE

OF THE

PHÆNOGAMOUS AND VASCULAR CRYPTOGAMOUS

PLANTS OF WORCESTER COUNTY,

MASSACHUSETTS.

BY JOSEPH JACKSON.

SECOND EDITION, REVISED AND ENLARGED.

WORCESTER :

1894.

WORCESTER :

PRIVATE PRESS OF FRANKLIN P. RICE.

MDCCCXCIV.

PREFACE.

In the autumn of 1883 I prepared for publication the first *Catalogue of the Plants of Worcester County, Massachusetts*, containing eight hundred and twelve species and well-marked varieties. In the meantime, by the kind and generous help of many friends, all of whom will understand, I hope, that whether their names appear in these pages or not, my sense of obligation to them is here fully acknowledged and gratefully remembered, more than two hundred additions have been made.

This Catalogue, revised and enlarged, is now issued in the hope that it may serve as a basis for still further additions, so that, in due time, our county may have its flora as fully recorded as possible. Its main purpose is to help and to encourage the beginner and to afford pleasure to those who take an interest in untamed and unpruned Nature.

Much yet remains to be done. The mosses, lichens, fungi and algæ furnish a wide field for the coöperation of workers in different parts of the county for some years. The yearly increasing interest in such work and the greatly increased facilities for prosecuting it are hopeful signs of its accomplishment.

The arrangement and the nomenclature of this Catalogue are those of Gray's *Manual of the Botany of the Northern United States*,—Sixth edition, 1890—the book most readily accessible to all persons interested in the study of our local flora. In the present state of botanical nomenclature, and for the purpose for which this Catalogue is intended, I have not thought it wise or necessary to adopt proposed changes not yet generally accepted. The student interested in such matters can easily adapt himself to changing conditions.

Introduced species, so far as I have been able to determine the fact of introduction, are indicated by a *.

FLORA OF WORCESTER COUNTY.

Series I. PHÆNOGAMIA.

Class I. DICOTYLEDONEÆ.

Sub-class I. ANGIOSPERMÆ.

Division I. POLYPETALÆ.

RANUNCULACEÆ.

CLEMATIS, L.   Virgin's Bower.
  Virginiana, L.   Common.
ANEMONE, Tourn.   Anemone.
  cylindrica, Gray.   Long-fruited Anemone.
  Virginiana, L.   Common.
  nemorosa, L.   Anemone.   Very common.
HEPATICA, Dill.   Liver-leaf.   Hepatica.
  triloba, Chaix.   Common.
ANEMONELLA, Spach.
  thalictroides, Spach.   Rue-Anemone.   Not rare.

THALICTRUM, Tourn.   Meadow-Rue.
 dioicum, L.   Early Meadow-Rue.   Rocky woods.
 polygamum, Muhl.   Tall Meadow-Rue.   Common.
 purpurascens, L.   Purplish Meadow-Rue.   Grafton.
RANUNCULUS, Tourn.   Crowfoot.   Buttercup.
 aquatilis, L., var. trichophyllus, Gray.   White Water-
   Crowfoot.   Worcester, Gardner.
 multifidus, Pursh.   Yellow Water-Crowfoot.
 Flammula, L., var. reptans, E. Meyer.   Lancaster.
                F. L. Palmer.
 abortivus, L.   Small-flowered Crowfoot.   Common.
 recurvatus, Poir.   Hooked Crowfoot.   Woods, common.
 fascicularis, Muhl.   Early Buttercup.   Dry hills.
  "   "   var.   Sunnyside, Worcester.
                Prof. Eaton.
 septentrionalis, Poir.   Worcester.
 Pennsylvanicus, L. f.   Bristly Buttercup.   Barre.
              Miss Sara Lane.
 *bulbosus, L.   Bulbous Buttercup.   Common.
 *acris, L.   Tall Buttercup.   Very common.
CALTHA, L.   Marsh Marigold.
 palustris, L.   Cowslip, but not of literature.   Common.
COPTIS, Salisb.   Goldthread.
 trifolia, Salisb.   Three-leaved Goldthread.   Common.
AQUILEGIA, Tourn.   Columbine.
 Canadensis, L.   Wild Columbine.   Common.
ACTÆA, L.   Baneberry.
 spicata, L., var. rubra, Ait.   Red Baneberry.   In copses.
 alba, Bigel.   White Baneberry.   With the last.

MAGNOLIACEÆ.

LIRIODENDRON, L. Tulip-Tree.
    Tulipifera, L. Rare. *G. E. Stone.*

BERBERIDACEÆ.

BERBERIS, L. Barberry.
    *vulgaris, L. In eastern and southern parts of the county.
CAULOPHYLLUM, Michx. Blue Cohosh.
    thalictroides, Michx. Pappoose-root. Worcester,
      Princeton.
PODOPHYLLUM, L. May-Apple. Mandrake.
    peltatum, L. West Boylston. *Arba Pierce.*

NYMPHEACEÆ.

BRASENIA, Schreber. Water-Shield.
    peltata, Pursh. Common.
NYMPHEA, Tourn. Water-Lily.
    odorata, Ait. Sweet-scented Water-Lily. Common.
    "   "   var. minor, Sims. Rare.
NUPHAR, Smith. Yellow Pond-Lily.
    advena, Ait. f. Common.
    Kalmianum, Ait. Southbridge. *L. E. Ammidown.*

SARRACENIACEÆ.

SARRACENIA, Tourn. Side-saddle Flower.
    purpurea, L. Pitcher-Plant. Common.

PAPAVERACEÆ.

SANGUINARIA, Dill. Blood-root.
    Canadensis, L. Common on rocky hillsides.

CHELIDONIUM, L.   Celandine.
   *majus, L. Celandine.   Waste grounds.
PAPAVER, Tourn.   Poppy.
   *somniferum, L. Common Poppy. Sutton.
                                      Dr. G. C. Webber.

FUMARIACEÆ.

ADLUMIA, Raf.   Climbing Fumitory.
   cirrhosa, Raf.   Rare.
DICENTRA, Borkh.   Dutchman's Breeches.
   Cucullaria, DC. Dutchman's Breeches. Barre.
                                      Miss Sara Lane.
   Canadensis, DC. Squirrel Corn.   Barre, Winchendon.
CORYDALIS, Vent.   Corydalis.
   glauca, Pursh. Pale Corydalis.   In dry woods.
   aurea, Willd. Golden Corydalis.   Winchendon.
                                      F. R. Hathaway.
FUMARIA, Tourn.   Fumitory.
   *officinalis, L. Common Fumitory.   Waste places.

CRUCIFERÆ.

DENTARIA, Tourn.   Pepper-root.
   diphylla, L.   North Brookfield.   *Miss E. M. Reed.*
CARDAMINE, Tourn.   Bitter Cress.
   rhomboidea, DC. Spring Cress.   Wet meadows.
   hirsuta, L. Small Bitter Cress.   Wet places.
   parviflora.   A form answering to this is found on Mt.
      Wachusett.
ARABIS, L.   Rock Cress.
   Canadensis, L. Sickle-pod.   Sutton.
   perfoliata, Lam. Tower Mustard.   Sutton, Worcester.

ALYSSUM, Tourn.　　Alyssum.
　　*calycinum, L.　　　　　　　　　　　　　　　Occasional.
CAMELINA, Crantz.　　False Flax.
　　*sativa, Crantz.　　　　　　　　　　　　　　Occasional.
NASTURTIUM, R. Br.　　Water-Cress.
　　*officinale, R. Br.　True Water-Cress.　　　Common.
　　palustre, DC.　Marsh Cress.　　　　　　　　"
　　*　"　　"　var. hispidum, DC.　　　　Not rare.
　　*Armoracia, Fries.　Horseradish.　　　　　Escaped.
BARBAREA, R. Br.　　Winter Cress.
　　vulgaris, R. Br.　Common Winter Cress.
SISYMBRIUM, Tourn.　　Hedge Mustard.
　　*officinale, Scop.　Hedge Mustard.　　　Worcester.
　　　　　　　　　　　　　　　　　　　　　　　Prof. Eaton.
BRASSICA, Tourn.　　Mustard.
　　*nigra, Koch.　Black Mustard.　　　　　　Waste places.
CAPSELLA, Medic.　　Shepherd's Purse.
　　*Bursa-pastoris, Mœnch.　　　　　　　　Very common.
THLASPI, Tourn.　　Pennycress.
　　*arvense, L.　Field Pennycress.　Waste places.　Rare.
LEPIDIUM, Tourn.　　Peppergrass.
　　Virginicum, L.　Wild Peppergrass.　　　　Common.
　　*ruderale, L.　　　　　　　　　　　　　　Not common.
RAPHANUS, Tourn.　　Radish.
　　*Raphanistrum, L.　Wild Radish.　　Lunenburg.
　　　　　　　　　　　　　　　　　　　　　　　Mr. Kilburn.
　　*sativus, L.　Garden Radish.　Occasionally spontaneous.

CISTACEÆ.

HELIANTHEMUM, Tourn.　　Rock-rose.
　　Canadense, Michx.　Frost-weed.　Common in dry grounds.

LECHEA, Kalm.   Pinweed.

major, Michx.   Templeton.   *V. P. Parkhurst.*
minor, L.   Dry and sterile grounds.
tenuifolia, Michx.   With the last.

VIOLACEÆ.

VIOLA, Tourn.   Violet.
pedata, L.   Bird-foot Violet.   Common in sandy soil.
palmata, L.   Common Blue Violet.   Common.
"   " var. cucullata, Gray.   Low grounds, common.
sagittata, Ait..   Arrow-leaved Violet.   Dry grounds,
    common.
blanda, Willd.   Sweet White Violet.   Common in wet
    grounds.
primulæfolia, L.   Worcester.   *Prof. Eaton.*
lanceolata, L.   Lance-leaved Violet.   Wet grounds,
    common.
rotundifolia, Michx.   Early Yellow Violet.   Cold woods.
pubescens, Ait.   Downy Yellow Violet.   Common.
striata, Ait.   Pale Violet.   Leicester.
rostrata, Pursh.   Long-spurred Violet.   Fitchburg.
canina, L., var. Muhlenbergii, Gray. Dog Violet. Common.
*tricolor, L.   Pansy.   Occasionally spontaneous.

CARYOPHYLLACEÆ.

DIANTHUS, L.   Pink.
*Armeria, L.   Deptford Pink.   Dry pastures.
*deltoides, L.   Maiden Pink.   Rare.
*barbatus, L.   Sweet William.   Princeton.

SAPONARIA, L.   Soapwort.
  *officinalis, L.   Soapwort.   Naturalized.
SILENE, L.   Catchfly.
  *Cucubalus, Wibel.   Bladder Campion.   Meadows, not
      rare.
  Pennsylvanica, Michx.   Wild Pink.   Not abundant.
  antirrhina, L.   Sleepy Catchfly.   "   "
  *noctiflora, L.   Night-flowering Catchfly.   Rare.
                                        Dr. G. C. Webber.
LYCHNIS, Tourn.   Cockle.
  *vespertina, Sibth.   Evening Lychnis.   Occasional.
  *Githago, Lam.   Corn Cockle.   "
  *Flos-cuculi, L.   Ragged Robin.   "
ARENARIA, L.   Sandwort.
  Groenlandica, Spreng.   Mountain Sandwort.   Ashburn-
      ham.   *Prof. Vose.*
  lateriflora, L.   Sandwort.   Not rare.
STELLARIA, L.   Chickweed.
  *media, Smith.   Common Chickweed.   Very common.
  longifolia, Muhl.   Long-leaved Stitchwort.   Common.
  borealis, Bigel.   Northern Stitchwort.   In shaded places.
CERASTIUM, L.   Mouse-ear Chickweed.
  *viscosum, L.   Mouse-ear Chickweed.   Not common.
  *vulgatum, L.   Larger Mouse-ear Chickweed.   Common.
  arvense, L.   Field Chickweed.   Dry places.
SAGINA, L.   Pearlwort.
  procumbens, L.   In shaded pastures.   Rare.
BUDA, Adans.   Sand-Spurrey.
  rubra, Dumort.   Common in dry soil.

SPERGULA, L.   Spurrey.
   *arvensis, L.   Corn Spurrey.   Not rare.

PORTULACACEÆ.

PORTULACA, Tourn.   Purslane.
   *oleracea, L.   Common Purslane.   Common in gardens.
CLAYTONIA, Gronov.   Spring-Beauty.
   Caroliniana, Michx.   Worcester, Holden.

HYPERICACEÆ.

HYPERICUM, Tourn.   St. John's-wort.
   ellipticum, Hook.   Wet places.   Common.
   *perforatum, L.   Common St. John's-wort.   Fields.
      Common.
   maculatum, Walt.   Common in damp places.
   mutilum, L.   In low grounds.
   Canadense, L.   Common in low grounds.
   nudicaule, Walt.   Pine-weed.   In dry grounds.
ELODES, Adans.   Marsh St. John's-wort.
   campanulata, Pursh.   Common in swampy grounds.

MALVACEÆ.

MALVA, L.   Mallow.
   *rotundifolia, L.   Common Mallow.   Common.
   *sylvestris, L.   High Mallow.   Waysides.
   *crispa, L.   Curled Mallow.   Sparingly escaped.
   *moschata, L.   Musk Mallow.   "   "

TILIACEÆ.

TILIA, Tourn.   Basswood.
   Americana, L.   Basswood.   Common.

LINACEÆ.

Linum, Tourn.   Flax.

*usitatissimum, L.   Common Flax.   Occasionally
spontaneous.

GERANIACEÆ.

Geranium, Tourn.   Cranesbill.

maculatum, L.   Wild Cranesbill.   Common.

Robertianum, L.   Herb Robert.   Mt. Wachusett,
Prof. Eaton; Worcester, *G. Coult.*

Carolinianum, L.   Cultivated grounds.

*pusillum, L.   Waste places.   Rare.

Erodium, L'Her.   Storksbill.

*cicutarium, L'Her.   Occasionally in meadows.

Oxalis, L.   Wood-Sorrel.

Acetosella, L.   Common Wood-Sorrel.   Northern part of
the county.

violacea, L.   Violet Wood-Sorrel.   Northern part of the
county.

corniculata, L., var. stricta, Sav.   Yellow Wood-Sorrel.
Common.

Impatiens, L.   Balsam.

pallida, Nutt.   Pale Touch-me-not.   Not rare.

fulva, Nutt.   Spotted Touch-me-not.   Common.

RUTACEÆ.

Xanthoxylum, L.   Prickly Ash.

Americanum, Mill.   Northern Prickly Ash.   Millbury.

ILICINEÆ.

Ilex, L.   Holly.
    verticillata, Gray.   Winterberry.   Common.
    lævigata, Gray.   Smooth Winterberry.   Swamps.
Nemopanthes, Raf.   Mountain Holly.
    fascicularis, Raf.   Swampy woods.

CELASTRACEÆ.

Celastrus, L.   Bitter-sweet.
    scandens, L.   Climbing Bitter-sweet.   Common.

RHAMNACEÆ.

Rhamnus, Tourn.   Buckthorn.
    *cathartica, L.   Common Buckthorn.   Naturalized.
Ceanothus, L.   New Jersey Tea.
    Americanus, L.   New Jersey Tea.   Common.

VITACEÆ.

Vitis, Tourn.   Grape.
    Labrusca, L.   Northern Fox-Grape.   Common.
    cordifolia, Michx.   Frost Grape.   Not uncommon.
    riparia, Michx.   G. B. Emerson.
Ampelopsis, Michx.   Virginian Creeper.
    quinquefolia, Michx.   Common.

SAPINDACEÆ.

Acer, Tourn.   Maple.
    Pennsylvanicum, L.   Striped Maple.   Common.
    spicatum, Lam.   Mountain Maple.   Northern part of
      the county.

saccharinum, Wang. Sugar Maple.   Common.
dasycarpum, Ehrh. White Maple.   Northern part of
          the county.
rubrum, L. Red Maple.   Common.

ANACARDIACEÆ.

RHUS, L.   Sumach.
   typhina, L. Staghorn Sumach.   Common.
   glabra, L. Smooth Sumach.   "
   copallina, L. Dwarf Sumach.   "
   venenata, DC. Poison Dogwood.   "
   Toxicodendron, L. Poison Ivy.   "
        "   " var. radicans, L.   "

POLYGALACEÆ.

POLYGALA, Tourn.   Milkwort.
   paucifolia, Willd. Fringed Polygala.   Common.
   polygama, Walt.   Dry sandy soil. Auburn.
   sanguinea, L.   Common in moist places.
   Nuttallii, Torr. & Gray.   Dry soil. Not common.
   verticillata, L.   Dry soil, common.

LEGUMINOSÆ.

BAPTISIA, Vent.   False Indigo.
   tinctoria, R. Br. Wild Indigo.   Abundant.
GENISTA, L.   Woad-Waxen.
   *tinctoria, L. Dyer's Green-weed.   Rare.
LUPINUS, Tourn.   Lupine.
   perennis, L. Wild Lupine.   Quite common.

TRIFOLIUM, Tourn.   Clover.

*arvense, L.   Rabbit-foot Clover.   Old fields, common.

*pratense, L.   Red Clover.   Common in meadows.

*repens, L.   White Clover.   "   "   "

*hybridum, L.   Alsike Clover.   Worcester.

*agrarium, L.   Yellow Clover.   "

*procumbens, L.   Low Hop-Clover.   Millbury.

MELILOTUS, Tourn.   Melilot.

*officinalis, Willd.   Yellow Melilot.   Not common.

*alba, Lam.   White Melilot.   Common.

MEDICAGO, Tourn.   Medick.

*sativa, L.   Alfalfa.   Spreading.

*lupulina, L.   Black Medick.   Waste places.

*maculata, Willd.   Spotted Medick.   "   "

*denticulata.   "   "

TEPHROSIA, Pers.   Hoary Pea.

Virginiana, Pers.   Goat's Rue.   Dry soil, not rare.

ROBINIA, L.   Locust-tree.

Pseudacacia, L.   Common Locust.   Naturalized.

viscosa, L.   Clammy Locust.   "

CORONILLA, L.

*varia, L.   Rare.

DESMODIUM, Desv.   Tick-Trefoil.

nudiflorum, DC.   Dry woods, not abundant.

acuminatum, DC.   Rich woods.

rotundifolium, DC.   Dry woods, not rare.

canescens, DC.   Southbridge. *L. E. Ammidown.*

paniculatum, DC.   Copses, common.

Canadense, DC.   Common on roadsides.

rigidum, DC.	Common.
ciliare, DC.	Dry hills, not uncommon.
Marilandicum, F. Boott.	Copses.
LESPEDEZA, Michx. Bush-Clover.	
procumbens, Michx.	In dry soil.
violacea, Pers.	In dry copses.
Stuvei, Nutt.	" " "
polystachya, Michx.	Dry hills.
capitata, Michx.	With the last.
VICIA, Tourn. Vetch.	
*sativa, L. Common Vetch.	Waste places.
*hirsuta, Koch.	Barre. *Miss Sara Lane.*
Cracca, L.	Not abundant.
LATHYRUS, Tourn. Vetchling.	
ochroleucus, Hook.	Barre. *Miss Sara Lane.*
APIOS, Boerh. Ground-nut.	
tuberosa, Moench.	Low grounds, common.
PHASEOLUS, Tourn. Kidney Bean.	
perennis, Walt. Wild Bean.	Not common.
AMPHICARPÆA, Ell. Hog Pea-nut.	
monoica, Nutt.	Common in copses.
CASSIA, Tourn. Senna.	
Marilandica, L. Wild Senna.	Not common.
Chamæcrista, L. Partridge Pea.	Southern part of the county.
nictitans, L. Wild Sensitive-Plant.	With the last.

ROSACEÆ.

PRUNUS, Tourn. Plum. Cherry.	
pumila, L. Dwarf Cherry. Winchendon. *A. S. Allen.*	

Pennsylvanica, L. f. Wild Red Cherry. Common in woods.

Virginiana, L.   Choke-Cherry.   Copses.

serotina, Ehrh. Wild Black Cherry.   With the last.

SPIRÆA, L.   Meadow-Sweet.

salicifolia, L.   Common Meadow-Sweet.   Pastures, etc.

tomentosa, L.   Hardhack.   Low grounds.

PHYSOCARPUS, Maxim.   Nine-bark.

opulifolius, Maxim.   Worcester. *Miss E. J. Seaver.*

RUBUS, Tourn.   Bramble.

odoratus, L.   Purple Flowering-Raspberry.   Northern
   part of the county.

triflorus, Richardson. Dwarf Raspberry. Low grounds.

strigosus, Michx.   Red Raspberry.   Copses.

occidentalis, L.   Black Raspberry.   "

villosus, Ait.   High Blackberry.   "

Canadensis, L.   Low Blackberry.   Dry fields.

hispidus, L.   Running Swamp-Blackberry. Low grounds.

DALIBARDA, L.   Dalibarda.

repens, L.   Northern part of the county.

GEUM, L.   Avens.

album, Gmelin.   Borders of woods.

Virginianum, L.   "   "   "

strictum, Ait.   Moist meadows.

rivale, L. Purple Avens.   Moist meadows, common.

WALDSTEINIA, Willd.

fragarioides, Tratt. Barren Strawberry.   Winchendon.
                                    F. R. Hathaway.

FRAGARIA, Tourn.   Strawberry.

Virginiana, Mill.   Fields and woodlands, common.

vesca, L.   Less common than the last.

POTENTILLA, L.   Cinque-foil.

    Norvegica, L.   Pastures.

    argentea, L.   Silvery Cinque-foil.   Common.

    palustris, Scop.   Marsh Five-Finger.   Rare.

    fructicosa, L.   Shrubby Cinque-foil.   Eastern part of the county.

    tridentata, Ait.   Three-toothed Cinque-foil.   Mt. Wachusett.

    Canadensis, L.   Common Cinque-foil.   Common.

AGRIMONIA, Tourn.   Agrimony.

    Eupatoria, L.   Common Agrimony.   Copses.

POTERIUM, L.   Burnet.

    Canadense, Benth. & Hook.   Canadian Burnet.

ROSA, Tourn.   Rose.

    blanda, Ait.   Wild Rose.   Not common.

    Carolina, L.   "   "   Borders of swamps.

    lucida, Ehrh.   "   "   "   "   "

    humilis, Marsh.   Princeton.   *W. W. Bailey & J. F. Collins.*

    *rubiginosa, L.   Sweetbrier.   Pastures.

PYRUS, L.   Pear.   Apple.

    *Malus, L.   Apple   Occasionally self-sown.

    *communis, L.   Pear.   "   "   "

    arbutifolia, L. f.   Choke-berry.   In huckleberry pastures.

       "   "   var. melanocarpa, Hook.   Common.

    Americana, DC.   American Mountain-Ash.   Mt. Wachusett.

    *aucuparia, Gaert.   European Mountain-Ash.   Henshaw Pond, Leicester, spontaneous.

CRATÆGUS, L.   Hawthorn.

coccinea, L.   Scarlet-Thorn.   Common.

"   " var. mollis, Torr. & Gray.   Fitchburg. Rare.

punctata, Jacq.   Worcester.

AMELANCHIER, Medic.   June-berry.

Canadensis, Torr. & Gray.   Shad-bush.   Common.

SAXIFRAGACEÆ.

SAXIFRAGA, L.   Saxifrage.

Virginiensis, Michx.   Early Saxifrage.   Common.

Pennsylvanica, L.   Swamp Saxifrage.   "

TIARELLA, L.   False Mitre-wort.

cordifolia, L.   Abundant in northern part of the county.

MITELLA, Tourn.   Mitre-wort.

diphylla, L.   Common.

CHRYSOSPLENIUM, Tourn.   Golden Saxifrage.

Americanum, Schwein.   Common in wet places.

PARNASSIA, Tourn.   Grass of Parnassus.

Caroliniana, Michx.   In low grounds.

HYDRANGEA, Gronov.   Hydrangea.

arborescens, L.   Wild Hydrangea.   Barre.   *Miss Sara Lane.* Spontaneous.

RIBES, L.   Currant.   Gooseberry.

Cynosbati, L.   Wild Gooseberry.   Rocky woods.

rotundifolium, Michx.   Wild Gooseberry.   Princeton.

oxyacanthoides, L.   Wild Gooseberry.   Copses.

prostratum, L'Her.   Fetid Currant.   Mt. Wachusett.

floridum, L'Her.   Wild Black Currant.   Tatnuck.

rubrum, L., var. subglandulosum, Maxim.   Red Currant. Winchendon.

CRASSULACEÆ.

Penthorum, Gronov.   Ditch Stone-crop.
    sedoides, L.   Common in low lands.
Sedum, Tourn.   Stone-crop.
    ternatum, Michx.   Worcester. *Prof. Eaton.*
    *acre, L.   Mossy Stone-crop.   Established.
    *Telephium, L.   Live-for-ever.   "

DROSERACEÆ.

Drosera, L.   Sundew.
    rotundifolia, L.   Round-leaved Sundew. Bogs, common.
    intermedia, Hayne, var. Americana, DC.   "   "

HAMAMELIDEÆ.

Hamamelis, L.   Witch-Hazel.
    Virginiana, L.   Copses, common.

HALORAGEÆ.

Myriophyllum, Vaill.   Water-Milfoil.
    spicatum, L.   Ponds.
    verticillatum, L.   Ponds, common.
Proserpinaca, L.   Mermaid-weed.
    palustris, L.   Brooks. Not rare.
Callitriche, L.   Water-Starwort.
    verna, L.   In brooks. Millbury.

MELASTOMACEÆ.

Rhexia, L.   Meadow-Beauty.
    Virginica, L.   Sutton. *Dr. G. C. Webber.*

LYTHRACEÆ.

LYTHRUM, L.   Loosestrife.
*Salicaria, L.   Spiked Loosestrife.   Not common.
DECODON, Gmel.   Swamp Loosestrife.
  verticillatus, Ell.   Swamps.   Not common.

ONAGRACEÆ.

LUDWIGIA, L.   False Loosestrife.
  alternifolia, L.   Seed-box.   Swamps.
  palustris, Ell.   Water Purslane.   Swamps, common.
EPILOBIUM, L.   Willow-herb.
  angustifolium, L.   Fire-weed.   Common in clearings.
  lineare, Muhl.   In low lands.
  strictum, Muhl.   "   "   "
  coloratum, Muhl.   In low lands, common.
  adenocaulon, Haussk.   Wet grounds.
OENOTHERA, L.   Evening Primrose.
  biennis, L.   Common Evening Primrose.   Common.
    "   "   var. grandiflora, Lindl.   Rare.
  pumila, L.   Dry fields.   Quite common.
  fruticosa, L.   Sundrops.
GAURA, L.   Gaura.
  biennis, L.   Southbridge.   L. E. Ammidown.
CIRCÆA, Tourn.   Enchanter's Nightshade.
  Lutetiana, L.   Common in thickets.
  alpina, L.   Sutton, Spencer.   Miss A. E. Tucker.

CUCURBITACEÆ.

SICYOS, L.   One-seeded Bur-Cucumber.
  angulatus, L.   Thickets.

ECHINOCYSTIS, Torr. & Gray.   Wild Balsam-apple.
    lobata, Torr. & Gray.   Rich soil.

FICOIDEÆ.

MOLLUGO, L.   Indian-Chickweed.
    *verticillata, L.   Carpet-weed.   Common.

UMBELLIFERÆ.

DAUCUS, Tourn.   Carrot.
    *Carota, L.   Common.
ANGELICA, L.   Angelica.
    atropurpurea, L.   In low grounds.
HERACLEUM, L.   Cow-Parsnip.
    lanatum, Michx.   In low grounds.
PASTINACA, L.   Parsnip.
    *sativa, L.   Common.
THASPIUM, Nutt.   Meadow-Parsnip.
    aureum, Nutt.   Common in low meadows.
CRYPTOTÆNIA, DC.   Honewort.
    Canadensis, DC.   Worcester, Southbridge.
SIUM, Tourn.   Water Parsnip.
    cicutæfolium, Gmelin.   In wet places.
    Carsonii, Durand.   Swamps.   Sutton.
ZIZIA, Koch.   Zizia.
    aurea, Koch.   Upland meadows.
CARUM, L.   Caraway.
    *Carui, L.   Caraway.   Common.
CICUTA, L.   Water-Hemlock.
    maculata, L.   Musquash Root.   Common in swamps.
    bulbifera, L.   Worcester. *H. H. Kingsbury.*

Conium, L.   Poison Hemlock.
*maculatum, L.   Not rare.
Osmorrhiza, Raf.   Sweet Cicely.
  brevistylis, DC.   Worcester.   *Prof. Eaton.*
  longistylis, DC.   "   "   "
Hydrocotyle, Tourn.   Water Pennywort.
  Americana, L.   Common in damp woods.
Sanicula, Tourn.   Sanicle.
  Marylandica, L.   Common in copses.
    "   " var. Canadensis, Torr.   Worcester.

ARALIACEÆ.

Aralia, Tourn.   Ginseng.
  racemosa, L.   Spikenard.   Occasional.
  hispida, Vent.   Wild Elder.   Common in clearings.
  nudicaulis, L.   Wild Sarsaparilla.   Common.
  quinquefolia, Deesne. & Planch. Ginseng.   Not common.
  trifolia, Deesne. & Planch.   Dwarf Ginseng.   Common.

CORNACEÆ.

Cornus, Tourn.   Cornel.   Dogwood.
  Canadensis, L.   Bunch-berry.   Common northwards.
  florida, L.   Flowering Dogwood.   "   southwards.
  circinata, L'Her.   Round-leaved Cornel.   Not common.
  sericea, L.   Silky Cornel.   Common.
  stolonifera, Michx.   Red-osier Dogwood.   Sutton.
  paniculata, L'Her.   Panicled Cornel.   Common.
  alternifolia, L. f.   Alternate-leaved Cornel.   "
Nyssa, L.   Tupelo.
  sylvatica, Marsh.   Tupelo.   Occasional.

Division II. GAMOPETALÆ.

CAPRIFOLIACEÆ.

Sambucus, Tourn.   Elder.

    Canadensis, L. Common Elder.   Common in rich soil.

    racemosa, L. Red-berried Elder.   Rocky woods.

Viburnum, L.   Arrow-wood.

    lantanoides, Michx. Hobble-bush.   Common in the northern part of the county.

    Opulus, L. Cranberry-tree.   Rare in the southern part of the county.

    acerifolium, L. Maple-leaved Viburnum.   Common.

    dentatum, L. Arrow-wood.   Common in wet grounds.

    cassinoides, L. Withe-rod.   Swamps, common.

    Lentago, L. Sweet Viburnum.   Occasional.

Triosteum, L.   Fever-wort.

    perfoliatum, L.   Boylston.

Linnæa, Gronov.   Twin-flower.

    borealis, L.   Gardner, Templeton.

Symphoricarpos, Dill.   Snowberry.

    racemosus, Michx. Snowberry.   Farnumsville. *Miss K. I. Fish.*

Lonicera, L.   Honeysuckle.

    ciliata, Muhl. Fly-Honeysuckle.   Common.

    cærulea, L. Mountain Fly-Honeysuckle.   Boylston.

    hirsuta, Eaton. Hairy Honeysuckle.   *Bigelow's Fl. Bost.*

    glauca, Hill.   West Boylston, North Brookfield.

Diervilla, Tourn.   Bush-Honeysuckle.

    trifida, Moench.   Common in copses.

RUBIACEÆ.

Houstonia, L.   Houstonia.
    cærulea, L.. Bluets.   Common.
Cephalanthus, L.   Button-bush.
    occidentalis, L.   Button-bush.   Common in swamps.
Mitchella, L.   Partridge-berry.
    repens, L.   Partridge-berry.   Common.
Galium, L.   Bedstraw.
    *Mollugo, L.   Princeton.   *W. W. B. & J. F. C.*
    circæzans, Michx.   Wild Liquorice.   Rich woods.
    lanceolatum, Torr.   Wild Liquorice.   Dry woods.
      Worcester.   *Miss Sargent.*
    trifidum, L.   Small Bedstraw.   In wet grounds.
    "   " var. pusillum, Gray.   "   "   "
    asprellum, Michx.   Rough Bedstraw.   "   "   "
    triflorum, Michx.   Sweet-scented Bedstraw.   Rich woods.

DIPSACEÆ.

Dipsacus, Tourn.   Teasel.
    *sylvestris, Mill.   Wild Teasel.   Worcester.

COMPOSITÆ.

Vernonia, Schreb.   Iron-weed.
    Noveboracensis, Willd.   In swampy meadows.
Mikania, Willd.   Climbing Hemp-weed.
    scandens, L.   Tatnuck. *Mr. Kinney.* Rare.
Eupatorium, Tourn.   Thoroughwort.
    purpureum, L.   Joe-Pye Weed.   Common.
    rotundifolium, L., var. ovatum, Torr.   Worcester. *G. Coult.*

perfoliatum, L. Thoroughwort. Common.

ageratoides, L. White Snake-root. Rich woods, common.

aromaticum, L. Southbridge. *L. E. Ammidown.*

LIATRIS, Schreb. Blazing-Star.

scariosa, Willd. Sterling. *Miss E. J. Seaver.*

SOLIDAGO, L. Golden-rod.

squarrosa, Muhl. Southbridge. *L. E. Ammidown.*

cæsia, L. Rich woodlands, common.

latifolia, L. " " "

bicolor, L. White Golden-rod. Common.

puberula, Nutt. Princeton. *W. W. B. & J. F. C.*

uliginosa, Nutt. Swamps.

speciosa, Nutt. Barre. *Miss Sara Lane.*

odora, Ait. Sweet Golden-rod. Copses.

patula, Muhl. Swamps, not rare.

rugosa, Mill. Very common.

ulmifolia, Muhl. Southbridge. *L. E. Ammidown.*

Elliottii, Torr. & Gray. Swamps.

neglecta, Torr. & Gray. "

arguta, Ait. Copses.

juncea, Ait. Copses and banks, common.

serotina, Ait. Copses and along fences, "

" " var. gigantea, Gray. Near the last.

Canadensis, L. Very common.

nemoralis, Ait. Dry sterile fields, "

lanceolata, L. Moist soil, "

tenuifolia, Pursh. Southbridge. *L. E. Ammidown.*

SERICOCARPUS, Nees. White-topped Aster.

conyzoides, Nees. Dry grounds, common.

solidagineus, Nees. Worcester, Boylston.

Aster, L. Aster.

corymbosus, Ait.	Woodlands, common.
macrophyllus, L.	Moist woods, "
Novæ-Angliæ, L.	Moist grounds, not rare.
patens, Ait.	Dry grounds, common.
undulatus, L.	Copses, "
cordifolius, L.	Dry woodlands, "
sagittifolius, Willd.	Pastures, etc., "
lævis, L.	Borders of woodlands, "
ericoides, L.	Southbridge. *L. E. Ammidown*.
multiflorus, Ait.	Dry sandy soil, common.
dumosus, L.	Copses, "
vimineus, Lam.	Waste grounds, "
diffusus, Ait.	" " "
Tradescanti, L.	Low " "
paniculatus, Lam.	Shady moist banks, "
salicifolius, Ait.	Pastures and low grounds, "
Novi-Belgii, L.	Moist grounds, "
prenanthoides, Muhl.	Grafton. *Misses Putnam*.
puniceus, L.	Low grounds, very common.
umbellatus, Mill.	Moist thickets, common.
linariifolius, L.	Dry grounds, "
ptarmicoides, Torr. and Gray.	Barre. *Miss Sara Lane*.
acuminatus, Michx.	Rich woodlands.

Erigeron, L. Fleabane.

Canadensis, L. Horse-weed.		Common.
annuus, Pers. Daisy Fleabane.		Waste places.
strigosus, Muhl. " "		" "
" " var. discoideus, Robbins.		Millbury.

✓ * bellidifolius, Muhl.   Robin's Plantain.   Common.
✓   Philadelphicus, L.   Common Fleabane.   "
ANTENNARIA, Gærtn.   Everlasting.
✓   plantaginifolia, Hook.   Pastures, common.
ANAPHALIS, DC.   Everlasting.
     margaritacea, Benth. & Hook.   Pearly Everlasting.
                                                    Common.
GNAPHALIUM, L.   Cudweed.
     polycephalum, Michx.   Common Everlasting.   Common.
     decurrens, Ives.   Everlasting.   "
     uliginosum, L.   Low Cudweed.   Low grounds.
     purpureum, L.   Purplish Cudweed.   "   "
INULA, L.   Elecampane.
     'Helenium, L.   Roadsides and low pastures.
AMBROSIA, Tourn.   Ragweed.
     artemisiæfolia, L.   Roman Wormwood.   Very common.
XANTHIUM, Tourn.   Cocklebur.
     *Strumarium, L.   Worcester.   J. Coulson.
HELIOPSIS, Pers.   Ox-eye.
     lævis, Pers.   Worcester.   G. Coult.   From the West.
RUDBECKIA, L.   Cone-flower.
     laciniata, L.   Leicester.   H. H. Kingsbury.
     hirta, L.   From the West.   Meadows, common.
HELIANTHUS, L.   Sunflower.
     annuus, L.   Common Sunflower.   Escaped.
     divaricatus, L.   Thickets, not rare.
     strumosus, L.   "   common.
     decapetalus, L.   Copses, not common.
     tuberosus, L.   Jerusalem Artichoke.   "   "

BIDENS, L. Bur-Marigold.

frondosa, L. Common Beggar-ticks. Waste grounds.

connata, Muhl. Swamp Beggar-ticks. Low "

cernua, L. Smaller Bur-Marigold. Wet places.

chrysanthemoides, Michx. Larger Bur-Marigold. Swamps.

Beckii, Torr. Water Marigold. Lake 'Quinsigamond.

GALINSOGA, Ruiz & Pavon. Galinsoga.

*parviflora, Cav. Worcester, Millbury.

ANTHEMIS, L. Chamomile.

*Cotula, DC. May-weed. Roadsides, common.

ACHILLEA, L. Yarrow.

Millefolium, L. Common Yarrow. Common.

CHRYSANTHEMUM, Tourn. Ox-eye Daisy.

*Leucanthemum, L. Ox-eye Daisy. Very common.

*Parthenium, Pers. Feverfew. Barre. *Miss Sara Lane.*

TANACETUM, L. Tansy.

*vulgare, L. Common Tansy. Quite common.

* " " var. crispum. *G. E. Stone.*

ARTEMISIA, L. Wormwood.

*vulgaris, L. Common Mugwort. Worcester.

Prof. Eaton.

*Absinthium, L. Wormwood. Worcester. " "

TUSSILAGO, Tourn. Coltsfoot.

*Farfara, L. Fitchburg. *E. A. Hartwell.*

SENECIO, Tourn. Groundsel.

aureus, L. Squaw-weed. Very common.

ERECHTITES, Raf. Fireweed.

hieracifolia, Raf. Common in burned clearings.

ARCTIUM, L. Burdock.

*Lappa, L., var. minus. Roadsides and waste places.

CNICUS, Tourn.   Thistle.

*lanceolatus, Hoffm.   Common Thistle.   Common.

horridulus, Pursh.   Southbridge.   *L. E. Ammidown.*

altissimus, Willd., var. discolor, Gray.   Not rare.

muticus, Pursh.   Swamp Thistle.   Worcester.   *Prof. Eaton.*

pumilus, Torr.   Pasture Thistle.   Southbridge.

L. E. Ammidown.

*arvensis, Hoffm.   Canada Thistle.   Common.

KRIGIA, Schreber.   Dwarf Dandelion.

Virginica, Willd.   Upland woods and pastures.

CICHORIUM, Tourn.   Chicory.

*Intybus, L.   Roadsides, not rare.

LEONTODON, L.   Fall Dandelion.

*autumnalis, L.   Meadows and roadsides.   Common
in southern part of the county.

HIERACIUM, Tourn.   Hawkweed.

*aurantiacum, L.   Winchendon.   *A. S. Allen.*

Canadense, Michx.   Copses, common.

paniculatum, L.   "   "

venosum, L.   Rattlesnake-weed.   In dry woods.

scabrum, Michx.   Dry open woods.

Gronovii, L.   Hairy Hawkweed.   Common.

PRENANTHES, Vaill.   Rattlesnake-root.

alba, L.   White Lettuce.   Rich woods, common.

serpentaria, Pursh.   Lion's-foot.   *G. E. Stone.*

altissima, L.   Rich woods, not rare.

TARAXACUM, Haller.   Dandelion.

*officinale, Weber.   Common Dandelion.   Everywhere.

LACTUCA, Tourn.   Lettuce.

Canadensis. L.   Wild Lettuce.   Rich soil, common.

hirsuta, Muhl.   Southbridge. *L. E. Ammidown.*
leucophæa, Gray.   Low grounds, not rare.
Sonchus, L.   Sow-Thistle.
  *oleraceus, L.   Common Sow-Thistle.   Waste places.
  *asper, Vill.   Spiny-leaved Sow-Thistle.   "   "
  *arvensis, L.   Field Sow-Thistle.   "   "

LOBELIACEÆ.

Lobelia, L.   Lobelia.
  cardinalis, L.   Cardinal-flower.   Low grounds.
  syphilitica, L.   Great Lobelia.   "   "
  spicata, Lam.   ·   Moist sandy soil.
  inflata, L.   Indian Tobacco.   Low grounds.
  Dortmanna, L.   Water Lobelia.   Borders of ponds.

CAMPANULACEÆ.

Specularia, Heister.   Venus' Looking-glass.
  perfoliata, A. DC.   Dry open grounds.
Campanula, Tourn.   Bellflower.
  *rapunculoides, L.   Naturalized.
  rotundifolia, L.   Harebell.   Winchendon. *A. S. Allen.*
  aparinoides, Pursh.   Marsh Bellflower.   Wet meadows.

ERICACEÆ.

Gaylussacia, HBK.   Huckleberry.
  dumosa, Torr. & Gray.   Dwarf Huckleberry.   North-
    borough. *Arba Pierce.*
  frondosa, Torr. & Gray.   Dangleberry. Copses, common.
  resinosa, Torr. & Gray.   Black Huckleberry.   Pastures.
    common.

Leucothoë racemosa, Gray.

The Leucothoë.

Thou hast few *rivals, rare Leucothoë*
In *grace and loveliness among the flowers.*

VACCINIUM, L.   Blueberry.   Cranberry.

Pennsylvanicum, Lam.   Dwarf Blueberry.   Common.

Canadense, Kalm.   Winchendon.   *A. S. Allen.*

vacillans, Solander.   Low Blueberry.   Common.

corymbosum, L.   High Blueberry.   "

"   "   var. atrococcum, Gray.   "

Oxycoccus, L.   Small Cranberry.   Bogs,   "

macrocarpon, Ait.   Large Cranberry.   "   "

CHIOGENES, Salisb.   Creeping Snowberry.

serpyllifolia, Salisb.   Swamps.   Auburn.

ARCTOSTAPHYLOS, Adans.   Bearberry.

Uva-ursi, Spreng.   Bare hills, not common.

EPIGÆA, L.   Trailing Arbutus.

repens, L.   Woods and pastures, common.

GAULTHERIA, Kalm.   Wintergreen.

procumbens, L.   Checkerberry.   Common.

ANDROMEDA, L.   Andromeda.

polifolia, L.   Formerly found in Westborough ; now
    in Whitehall Pond, just over the line into Middle-
    sex County.   It will undoubtedly be found again
    in this county.

ligustrina, Muhl.   Copses, common.

LEUCOTHOË, Don.   Leucothoë.

racemosa, Gray.   Millbury.   Not common.

CASSANDRA, Don.   Leather-Leaf.

calyculata. Don.   Swamps, common.

KALMIA, L.   American Laurel.

latifolia, L.   Mountain Laurel.   Common.

angustifolia, L.   Sheep Laurel.   "

glauca, Ait.   Pale Laurel.   Swamps, not rare.

RHODODENDRON, L.   Rose Bay.   Azalea.
  viscosum, Torr.   White Swamp-Honeysuckle.   Common.
  nudiflorum, Torr.   Swamp Pink.   "
  Rhodora, Don.   Rhodora.   Swamps,   "
  maximum, L.   Rhododendron.   Auburn, Sturbridge.
LEDUM, L.   Labrador Tea.
  latifolium, Ait.   Worcester, Hubbardston.
CLETHRA, Gronov.   White Alder.
  alnifolia, L.   Sweet Pepperbush.   Wet copses, common.
CHIMAPHILA, Pursh.   Pipsissewa.
  umbellata, Nutt.   Prince's Pine.   Dry woods, common.
  maculata, Pursh.   Spotted Wintergreen.   With the last.
MONESES, Salisb.   One-flowered Pyrola.
  grandiflora, Salisb.   Abundant in northern part of the
      county. F. L. Palmer.
PYROLA, Tourn.   Wintergreen.
  secunda, L.   Copses, not common.
  chlorantha, Swartz.   Woods, common.
  elliptica, Nutt.   Shin-leaf.   "
  rotundifolia, L.   Copses,   "
MONOTROPA, L.   Indian Pipe.
  uniflora, L.   Indian Pipe.   Rich woods, common.
  Hypopitys, L.   Pine-sap.   With the last.

PRIMULACEÆ.

TRIENTALIS, L.   Chickweed-Wintergreen.
  Americana, Pursh.   Star-flower.   Woods, common.
STEIRONEMA, Raf.   Loosestrife.
  ciliatum, Raf.   Low grounds, common.
  lanceolatum, Gray.   Low grounds, not   "

Ledum latifolium, Ait.

The Labrador Tea.

The northlands claim **with pride thy** blossoms fair,
And yet thou spurnest **not our summer** air.

LYSIMACHIA, Tourn.　Loosestrife.
　*vulgaris, L.　　　　　Bolton.　*Miss J. M. Nichols.*
　quadrifolia, L.　　　　　　Thickets, common.
　stricta, Ait.　　　　Low grounds,　"
　*nummularia, L.　Moneywort.　　　　Escaped.
　thyrsiflora, L.　Tufted Loosestrife.　Northborough.
　　　　　　　　　　　　　　　　Arba Pierce.
ANAGALLIS, Tourn.　Pimpernel.
　*arvensis, L.　　　Waste grounds.　Southbridge.

OLEACEÆ.

FRAXINUS, Tourn.　Ash.
　Americana, L.　White Ash.　　　Common.
　sambucifolia, Lam.　Black Ash.　　Not rare.

APOCYNACEÆ.

APOCYNUM, Tourn.　Dogbane.
　androsæmifolium, L.　Spreading Dogbane.　Not rare.
　cannabinum, L.　Indian Hemp.　　　"　　"

ASCLEPIADACEÆ.

ASCLEPIAS, L.　Milkweed.
　tuberosa, L.　Butterfly-weed.　　　　Millbury.
　purpurascens, L.　Purple Milkweed.　Roadsides.
　incarnata, L.　Swamp Milkweed.　　Swamps.
　　"　　" var. pulchra, Pers.　With the last.
　Cornuti, Decsne.　Common Milkweed.　Common.
　obtusifolia, Michx.　　　　Not common.
　phytolaccoides. Pursh.　Poke Milkweed.　Moist copses.
　quadrifolia, L.　　　　Millbury, Worcester.

ACERATES, Ell.   Green Milkweed.
viridiflora, Ell.   In the western part of the county.

GENTIANACE.E.

GENTIANA, Tourn.   Gentian.
crinita, Froel.   Fringed Gentian.   Common.
Andrewsii, Griseb.   Closed Gentian.   "
MENYANTHES, Tourn.   Buckbean.
trifoliata, L.   Sutton, Paxton, etc.
LIMNANTHEMUM, Gmelin.   Floating Heart.
lacunosum, Griseb.   Lake Quinsigamond. Crystal
Lake, Gardner. *F. L. Palmer.*

POLEMONIACE.E.

POLEMONIUM, Tourn.   Greek Valerian.
reptans, L.   Rare. *Miss G. Hakes.*

BORRAGINACEÆ.

CYNOGLOSSUM, Tourn.   Hound's-Tongue.
*officinale, L.   Common Hound's-Tongue.   Waste
grounds.
Virginicum, L.   Wild Comfrey.   Millbury, Princeton.
ECHINOSPERMUM, Lehm.   Stickseed.
Virginicum, Lehm.   Beggar's Lice.   Common.
MYOSOTIS, Dill.   Forget-me-not.
laxa, Lehm.   Spencer, Lunenburg.
LITHOSPERMUM, Tourn.   Gromwell.
*arvense, L.   Corn Gromwell.   Occasional.
*officinale, L.   Common Gromwell. *Miss E. J. Seaver.*

Symphytum, Tourn.   Comfrey.
    *officinale, L.   Common Comfrey.   Escaped.
Echium, Tourn.   Viper's Bugloss.
    *vulgare, L.   Blue-weed. Southbridge.   *L. E. Ammidown.*

CONVOLVULACEÆ.

Convolvulus, Tourn.   Bindweed.
    spithamæus, L.   Not rare.
    sepium, L.   Hedge Bindweed.   " "
    *arvensis, L.   Old fields.
Cuscuta, Tourn.   Dodder.
    Gronovii, Willd.   Wet places.   Common.

SOLANACEÆ.

Solanum, Tourn.   Nightshade.
    *Dulcamara, L.   Bittersweet.   Common.
    nigrum, L.   Common Nightshade.   Southbridge.
Physalis, L.   Ground Cherry.
    Virginiana, Mill.   Roadsides.   Not common.
Nicandra, Adans.   Apple of Peru.
    *physaloides, Gærtn.   Waste grounds.   *G. E. Stone.*
Datura, L.   Thorn-Apple.
    *Stramonium, L.   Waste grounds.   Not rare.
    *Tatula, L.   " "   " "

SCROPHULARIACEÆ.

Verbascum, L.   Mullein.
    *Thapsus, L.   Common Mullein.   Fields.   Common.
    *Blattaria.   Moth Mullein.   Roadsides.   Not common.

LINARIA, Tourn.   Toad-Flax.

Canadensis, Dumont.   Sandy soil.   Common.

*vulgaris, Mill.   Butter and Eggs.   "

CHELONE, Tourn.   Snake-head.

glabra, L.   Wet places.   Common.

PENSTEMON, Mitchell.   Beard-tongue.

pubescens, Solander.   Worcester.   *Miss E. F. Brown.*

MIMULUS, L.   Monkey-flower.

ringens, L.   Wet places.   Common.

GRATIOLA, L.   Hedge-Hyssop.

Virginiana, L.   Lake Quinsigamond.   *Prof. Eaton.*

aurea, Muhl.   Sandy swamps.   Common.

ILYSANTHES, Raf.   False Pimpernel.

riparia, Raf.   Wet places.

VERONICA, L.   Speedwell.

Anagallis, L.   Water Speedwell.

scutellata, L.   Marsh Speedwell.   Bogs.   Common.

officinalis, L.   Common Speedwell.   Dry hills.   "

*Chamædrys, L.   *G. E. Stone.*

serpyllifolia, L.   Thyme-leaved Speedwell.   Common.

*arvensis, L.   Corn Speedwell.   Rather rare.

*agrestis, L.   Field Speedwell.   Southbridge.

                                   L. E. Ammidown.

GERARDIA, L.   False Foxglove.

pedicularia, L.   Woods and thickets.

flava, L.   Downy False Foxglove.   Common in dry
     woods.

quercifolia, Pursh.   Smooth False Foxglove.   With the
     first.

purpurea, L.   Purple Gerardia.   In low grounds.
tenuifolia, Vahl.   Slender Gerardia.   Southbridge.
CASTILLEIA, Mutis.   Painted-Cup.
    coccinea, Spreng.   Scarlet Painted-Cup.   Occasional.
SCHWALBEA, Gronov.   Chaff-seed.
    Americana, L.   Southbridge.   *I. E. Ammidown.*
PEDICULARIS, Tourn.   Lousewort.
    Canadensis, L.   Common Lousewort.   Common.
  * lanceolata, Michx.   Rather rare.
MELAMPYRUM, Tourn.   Cow-Wheat.
    Americanum, Michx.   Open woods.   Common.

OROBANCHACEÆ.

EPIPHEGUS, Nutt.   Beech-drops.
    Virginiana, Bart.   Ashburnham.   *Prof. Vose.*
CONOPHOLIS, Wallroth.   Squaw-root.
    Americana, Wallroth.   Ashburnham.   *Prof. Vose.*
APHYLLON, Mitchell.   Naked Broom-rape.
    uniflorum, Gray.   One-flowered Cancer-root.   Common.

LENTIBULARIACEÆ.

UTRICULARIA, L.   Bladderwort.
    inflata, Walt.   Ponds.   Not rare.
    vulgaris, L.   Greater Bladderwort.   Ponds, common.
    minor, L.   Smaller   "   Ponds.   *G. E. Stone.*
    gibba, L.   Ponds.   *Miss E. J. Seaver.*
    intermedia, Hayne.   Pools.   *G. E. Stone.*
    purpurea, Walt.   Lake Quinsigamond.   "   "   "
    cornuta, Michx.   Lunenburg.   *F. L. Palmer.*

PEDALIACEÆ.

Martynia, L.   Unicorn-plant.
proboscidea, Glox.   Worcester. *Prof. Eaton.*

VERBENACEÆ.

Verbena, Tourn.   Vervain.
urticæfolia, L.   White Vervain.   Waste grounds.
hastata, L.   Blue Vervain.   "   "
Phryma, L.   Lopseed.
Leptostachya, L.   Open woods. Not abundant.

LABIATÆ.

Trichostema, L.   Blue Curls.
dichotomum, L.   Bastard Pennyroyal.   Sandy fields.
Collinsonia, L.   Horse-Balm.
Canadensis, L.   Stone-root.   Rich woods.
Mentha, Tourn.   Mint.
*viridis, L.   Spearmint.   Not rare.
*piperita, L.   Peppermint.   Common.
*aquatica, L.   Water Mint. Winchendon. *A. S. Allen.*
*arvensis, L.   Corn Mint. Southbridge. *L. E. Ammi-
    down.*
Canadensis, L.   Wild Mint.
Lycopus, Tourn.   Water Horehound.
Virginicus, L.   Bugle-weed.   In low grounds.
sessilifolius, Gray.   Not common.
sinuatus, Ell.   In low grounds. Quite common.
Pycnanthemum, Michx.   Mountain Mint.   *
lanceolatum, Pursh.   Spencer. *J. C. Lyford.*

linifolium, Pursh. *G. E. Stone.*

muticum, Pers. Mt. Wachusett. *Mrs. G. B. Stearns.*

THYMUS, Tourn. Thyme.

*Serpyllum, L. Creeping Thyme. Worcester.

CALAMINTHA, Tourn. Calamint.

Clinopodium, Benth. Basil. Millbury.

MELISSA, L. Balm.

*officinalis, L. Common Balm. Escaped. *G. E. Stone.*

HEDEOMA, Pers. Mock Pennyroyal.

pulegioides, Pers. American Pennyroyal. Not rare.

NEPETA, L. Cat-Mint.

*Cataria, L. Catnip. Near dwellings, common.

*Glechoma, Benth. Gill-over-the-ground. Quite common.

SCUTELLARIA, L. Skullcap.

lateriflora, L. Mad-dog Skullcap. Wet places, common.

galericulata, L. Wet places. With the last.

BRUNELLA, Tourn. Self-heal.

vulgaris, L. Common Self-heal. Common.

LEONURUS, L. Motherwort.

*Cardiaca, L. Waste places, common.

LAMIUM, L. Dead-Nettle.

*amplexicaule, L. Waste grounds. *Miss A. H. Tucker.*

*purpureum, L. Not common.

*maculatum, L. Escaped. *R. C. Manning.*

GALEOPSIS, L. Hemp-Nettle.

*Tetrahit, L. Waste places.

STACHYS, Tourn. Hedge-Nettle.

palustris, L. Wet grounds. Tatnuck. *Mr. Kinney.*

PLANTAGINACEÆ.

PLANTAGO, Tourn. Plantain.
  major, L. Common Plantain. Very common.
  Rugelii, Decaisne. Princeton. *W. W. B. & J. F. C.*
  *lanceolata, L. Ribgrass. Very common.
  Patagonica, Jacq., var. aristata, Gray. Southbridge.
  L. E. Ammidown.

DIVISION III. APETALÆ.

ILLECEBRACEÆ.

ANYCHIA, Michx. Forked Chickweed.
  dichotoma, Michx. Open grounds.
SCLERANTHUS, L. Knawel.
  *annuus, L. Worcester. *J. Coulson.*

AMARANTACEÆ.

AMARANTUS, Tourn. Amaranth.
  *hypochondriacus, L. Fitchburg. *Prof. Eaton.*
  *retroflexus, L. Cultivated grounds, common.
  albus, L. Tumble Weed. Waste grounds.

CHENOPODIACEÆ.

CHENOPODIUM, Tourn. Pigweed.
  *album, L. Very common.
  hybridum, L. Maple-leaved Goosefoot. Waste places.
  *Botrys, L. Jerusalem Oak. Lunenburg. *Mr. Kilburn.*

PHYTOLACCACEÆ.

PHYTOLACCA, Tourn. Pokeweed.
  decandra, L. Common Poke. Copses.

POLYGONACEÆ.

RUMEX, L.   Dock.

*crispus, L. Curled Dock.   Common.

*obtusifolius, L. Bitter Dock. Worcester. *Prof. Eaton.*

*conglomeratus, Murray. Smaller Green Dock.   Moist places.

*Acetosella, L. Field Sorrel.   Common.

POLYGONUM, Tourn.   Knotweed.

aviculare, L.   Common in yards and waste places.

erectum, L.   Waysides, common.

tenue, Michx.   Worcester. *Prof. Eaton.*

Pennsylvanicum, L.   Royalston.   "   "

amphibium, L.   North Worcester.

Muhlenbergii, Watson.   "   "

*orientale, L. Prince's Feather.   Escaped.

*Persicaria, L. Lady's Thumb.   *G. E. Stone.*

Hydropiper, L. Common Smartweed.   Common.

acre, HBK. Water Smartweed.   Wet places.

Virginianum, L.   Thickets.

arifolium, L. Halberd-leaved Tear-thumb. Low grounds.

sagittatum, L. Arrow-leaved Tear-thumb.   "   "

*Convolvulus, L. Black Bindweed.   Worcester.
                                        Prof. Eaton.

cilinode, Michx.   Low grounds. Quite common.

dumetorum, L., var. scandens, Gray.   Gardner.
                                        Prof. Eaton.

FAGOPYRUM, Tourn.   Buckwheat.

*esculentum, Mœnch.   Spontaneous. Occasional.

POLYGONELLA, Michx.   Joint-weed.

articulata, Meisn.   Mendon, Gardner.

PODOSTEMACEÆ.

PODOSTEMON, Michx.   River-weed.
    ceratophyllus, Michx.   Shallow streams.

ARISTOLOCHIACEÆ.

ASARUM, Tourn.   Wild Ginger.
    Canadense, L.   Millbury.   Not common.

LAURACEÆ.

SASSAFRAS, Nees.
    officinale, Nees.   Rich woods.   Common.
LINDERA, Thunb.   Wild Allspice.
    Benzoin, Blume.   Spice-bush.   Common.

THYMELÆACEÆ.

DIRCA, L.   Leatherwood.
    palustris, L.   Damp woods, common.
DAPHNE, Linn.   Mezereum.
    *Mezereum, L.   Southbridge.   *L. E. Ammidown.*

SANTALACEÆ.

COMANDRA, Nutt.   Bastard Toad-flax.
    umbellata, Nutt.   Dry grounds.   Common.

EUPHORBIACEÆ.

EUPHORBIA, L.   Spurge.
    maculata, L.   Roadsides, common.
    Preslii, Guss.   Millbury.
    corollata, L.   Southbridge.   *L. E. Ammidown.*

*Esula, L.   Rare. *Miss K. I. Fish.*
*Cyparissias, L.   Escaped. Common.
ACALYPHA, L.   Three-seeded Mercury.
    Virginica, L.   Fields and open places.

URTICACEÆ.

ULMUS, L.   Elm.
    fulva, Michx. Slippery Elm.   Rare.
    Americana, L.   American Elm.   Common.
CELTIS, Tourn.   Nettle-tree.
    occidentalis, L.   Hackberry.   Rare.
CANNABIS, Tourn.   Hemp.
    *sativa, L.   East Worcester.
HUMULUS, L.   Hop.
    Lupulus, L.   Common Hop.   Low grounds.
MORUS, Tourn.   Mulberry.
    *alba, L.   White Mulberry.   Spontaneous.
URTICA, Tourn.   Nettle.
    gracilis, Ait.   Waste places, common.
    *dioica, L.   Waste places. Rather rare.
LAPORTEA, Gaudichaud.   Wood-Nettle.
    Canadensis, Gaudichaud.   Rich soil. Rather rare.
PILEA, Lindl.   Richweed.
    pumila, Gray.   Moist places, common.
BŒHMERIA, Jacq.   False Nettle.
    cylindrica, Willd.   Rich soil. Quite common.

PLATANACEÆ.

PLATANUS, L.   Buttonwood.
    occidentalis, L.   Common.

JUGLANDACEÆ.

JUGLANS, L.   Walnut.
    cinerea, L.   Butternut.   Quite common.
CARYA, Nutt.   Hickory.
    alba, Nutt.   Shag-bark Hickory.   Common.
    porcina, Nutt. Pig-nut Hickory.   "
    amara, Nutt. Bitter-nut Hickory.   Not "

MYRICACEÆ.

MYRICA, L.   Bayberry.
    Gale, L.   Sweet Gale.   Wet borders of ponds.
    cerifera, L. Bayberry.   Dry sandy soil.
    asplenifolia, Endl. Sweet Fern.   Pastures, common.

CUPULIFERÆ.

BETULA, Tourn.   Birch.
    lenta, L.   Black Birch.   Common.
    lutea, Michx. f. Yellow Birch.   "
    populifolia, Ait. Gray   "   "
    papyrifera, Marshall. Paper Birch.   Rare in southern
        part of the county.
    nigra, L.   Red Birch.   Fitchburg. *E. A. Hartwell.*
ALNUS, Tourn.   Alder.
    incana, Willd. Speckled Alder.   Common.
    serrulata, Willd. Smooth   "   "
CORYLUS, Tourn.   Hazel-nut.
    Americana, Walt. Wild Hazel-nut.   Common.
    rostrata, Ait. Beaked Hazel-nut.   "
OSTRYA, Micheli. Hop-Hornbeam.
    Virginica, Willd.   Not rare.

Carpinus, L.   Hornbeam.
    Caroliniana, Walter.   Common.
Quercus, L.   Oak.
    alba, L.   White Oak.   Common.
    macrocarpa, Michx.   Bur Oak.   ·Worcester.
    bicolor, Willd.   Swamp White Oak.   Not rare.
    Prinus, L.   Chestnut-Oak.   Common.
        "   "   var. monticola, Michx.   "
    prinoides, Willd.   Not rare.
    rubra, L.   Red Oak.   Common.
    coccinea, Wang.   Scarlet Oak.   "
       "   "   var. tinctoria, Gray. Black Oak.   "
    ilicifolia, Wang.   Scrub-Oak.   "
Castanea, Tourn.   Chestnut.
    sativa, Mill., var. Americana, Michx.   Common.
Fagus, Tourn.   Beech.
    ferruginea, Ait.   Common.

SALICACEÆ.

Salix, Tourn.   Willow.
    nigra, Marsh.   Black Willow.   Not rare.
    lucida, Muhl.   Shining Willow.   Common.
    *fragilis, L.   Crack Willow.   Worcester. *Prof. Eaton.*
    *alba, L.   White Willow.   Not common.
    * "   "   var. vitellina, Koch.   Southbridge.
                           L. F. Ammidown.
    rostrata, Richardson.   Princeton. *W. W. B. & J. F. C.*
    discolor, Muhl.   Glaucous Willow.   Common.
    humilis, Marsh.   Prairie Willow.   Not   "
    tristis, Ait.   Dwarf Gray Willow.   Uxbridge.
                           Thos. Morong.

sericea, Marsh. Silky Willow.   Common.
cordata, Muhl. Heart-leaved Willow.   Worcester.
                                                Prof. Eaton.
myrtilloides, L.   Leicester.
POPULUS, Tourn.   Poplar.
    tremuloides, Michx. American Aspen.   Common.
    grandidentata, Michx. Large-toothed Aspen.   "
    balsamifera, L., var. candicans, Gray.   Balm of Gilead.
        Spontaneous.
    monilifera, Ait. Cotton-wood.   Rare.

CERATOPHYLLACEÆ.

CERATOPHYLLUM. L.   Hornwort.
    demersum, L.   Slow streams. *G. E. Stone.*
        "   " var. echinatum, Gray.   "   "   "

SUBCLASS II. GYMNOSPERMÆ.

CONIFERÆ.

PINUS, Tourn.   Pine.
    Strobus, L. White Pine.   Common.
    rigida, Mill. Pitch   "   "
    resinosa, Ait. Red   "   Templeton. *V. P. Parkhurst.*
PICEA, Link.   Spruce.
    nigra, Link. Black Spruce.   Not rare.
TSUGA, Carrière.   Hemlock.
    Canadensis, Carr.   Quite common.
ABIES, Link.   Fir.
    balsamea, Miller. Balsam Fir.   Northern part of the
        county.

LARIX, Tourn.   Larch.
    Americana, Michx.   Hackmatack.   Not rare.
CHAMÆCYPARIS, Spach.   White Cedar.
    sphæroidea, Spach.   Swamps, common.
JUNIPERUS, L.   Juniper.
    communis, L.   Common Juniper.   Not rare.
    Sabina, L., var. procumbens, Pursh.   "   "
    Virginiana, L.   Red Cedar.   "   "
TAXUS, Tourn.   Yew.
    Canadensis, Willd.   Worcester.   *G. Coult.*

CLASS II.   MONOCOTYLEDONEÆ.

HYDROCHARIDACEÆ.

ELODEA, Michx.   Water-weed.
    Canadensis, Michx.   Ponds, common.
VALLISNERIA, L.   Eel-grass.
    spiralis, L.   Ponds, common.

ORCHIDACEÆ.

MICROSTYLIS, Nutt.   Adder's Mouth.
    monophyllos, Lindl.   Spencer.   *H. H. Kingsbury.*
    ophioglossoides, Nutt.   Auburn.
LIPARIS, Richard.   Twayblade.
    liliifolia, Richard.   Millbury.   Rare.
    Lœselii, Richard.   Spencer.   *Miss A. E. Tucker.*
CORALLORHIZA, Haller.   Coral-root.
    innata, R. Br.   Swamps.   Millbury.

odontorhiza, Nutt.   Southbridge. *L. E. Ammidown.*
multiflora, Nutt.   Dry woods. Not rare.
SPIRANTHES, Richard.   Ladies' Tresses.
latifolia, Torr.   Southbridge. *L. E. Ammidown.*
cernua, Richard.   Wet places, common.
praecox, Watson.   Wet grassy places. Rare.
gracilis, Bigelow.   Pastures. Quite common.
simplex, Gray.   Winchendon. *A. S. Allen.*
GOODYERA, R. Br.   Rattlesnake-Plantain.
repens, R. Br.   Woods, mostly under evergreens.
pubescens, R. Br.   "   "   "   "
ARETHUSA, Gronov.   Arethusa.
bulbosa, L.   Bogs. Not rare.
CALOPOGON, R. Br.   Grass Pink.
pulchellus, R. Br.   Bogs. Not rare.
POGONIA, Juss.   Pogonia.
ophioglossoides, Nutt.   Common in southern part of
    the county.
verticillata, Nutt.   Woods. Rather rare.
ORCHIS, L.   Orchis.
spectabilis, L.   Rich woods. Rare.
HABENARIA, Willd.   Rein-Orchis.
tridentata, Hook.   Worcester. *G. Coult.*
virescens, Spreng.   Wet meadows. Millbury.
bracteata, R. Br.   Mt. Wachusett.
obtusata, Richardson.   Mt. Wachusett. *M. Pratt.*
Hookeri, Torr.   Damp woods. Not rare.
orbiculata, Torr.   Rich "   "   "
ciliaris, R. Br.   Northborough, *Dr. Bigelow;* Uxbridge,
    Miss Goldthwaite.

blephariglottis, Torr. Westminster. *Miss E. J. Seaver.*
leucophæa, Gray. Bolton. *Miss C. Knapp.*
Only one specimen was found, July, 1894.
lacera, R. Br. Ragged Fringed-Orchis. Quite common.
psycodes, Gray. Purple Fringed-Orchis. " "
fimbriata, R. Br. " " " " "

CYPRIPEDIUM, L. Lady's Slipper.
parviflorum, Salisb. Smaller Yellow Lady's Slipper. Rare.
pubescens, Willd. Larger " " " "
spectabile, Swartz. Showy Lady's Slipper. Ashburnham.
Prof. Vose.
acaule, Ait. Stemless Lady's Slipper. Common.

HÆMODORACEÆ.

ALETRIS, L. Colic-root.
farinosa, L. Grassy places. Not common.

IRIDACEÆ.

IRIS, Tourn. Flower-de-Luce.
versicolor, L. Larger Blue Flag. Common.
SISYRINCHIUM, L. Blue-eyed Grass.
angustifolium, Mill. Moist meadows, common.
anceps, Cav. Spencer. *H. H. Kingsbury.*

AMARYLLIDACEÆ.

HYPOXIS, L. Star-grass.
erecta, L. Meadows, common.

LILIACEÆ.

SMILAX, Tourn. Greenbrier.
herbacea, L. Carrion-Flower. Quite common.

rotundifolia, L.   Common Greenbrier.   Common.
ALLIUM, L.   Onion.
    tricoccum, Ait.   Wild Leek.   Tatnuck.
    Canadense, Kalm.   Wild Garlic.   Not rare.
ORNITHOGALUM, Tourn.   Star-of-Bethlehem.
    *umbellatum, L.   Escaped.
HEMEROCALLIS, L.   Day-Lily.
    *fulva, L.   Princeton. *W. W. B. & J. F. C.*
POLYGONATUM, Tourn.   Solomon's Seal.
    biflorum, Ell.   Woods, common.
ASPARAGUS, Tourn.   Asparagus.
    *officinalis, L.   Escaped.
SMILACINA, Desf.   False Solomon's Seal.
✓   racemosa, Desf.   False Spikenard.   Quite common.
    trifolia, Desf.   Bogs. Rare in southern part of the county.
MAIANTHEMUM, Wigg.   False Solomon's Seal.
✓   Canadense, Desf.   Moist woods, common.
STREPTOPUS, Michx.   Twisted-Stalk.
    amplexifolius, DC.   Mt. Wachusett. *M. Pratt.*
    roseus, Michx.   Woods, common.
CLINTONIA, Raf.   Clintonia.
✓   borealis, Raf.   Cold moist woods, common.
UVULARIA, L.   Bellwort.
/   perfoliata, L.   Rich woods, common.
OAKESIA, Watson.   Wild Oats.
✓   sessilifolia, Watson.   Low woods, common.
ERYTHRONIUM, L.   Dog's-tooth Violet.
✓   Americanum, Ker.   Rich moist grounds, common.
LILIUM, L.   Lily.
    Philadelphicum, L.   Wood Lily.   Not rare.

superbum, L. Turk's-cap Lily. Southbridge.

L. E. Ammidown.

Canadense, L. Wild Yellow Lily. Moist meadows. Common.

*tigrinum, Ker. Tiger Lily. Escaped.

MEDEOLA, Gronov. Indian Cucumber-root.

Virginiana, L. Rich woods, common.

TRILLIUM, L. Wake Robin.

erectum, L. Purple Trillium. Rich woods, common.

grandiflorum, Salisb. Barre. *Miss Sara Lane.* Rare.

cernuum, L. Wake Robin. Moist woods, common.

erythrocarpum, Michx. Painted Trillium. Common.

VERATRUM, Tourn. False Hellebore.

viride, Ait. Indian Poke. Common.

PONTEDERIACEÆ.

PONTEDERIA, L. Pickerel-weed.

cordata, L. Shallow water, common.

XYRIDACEÆ.

XYRIS, Gronov. Yellow-eyed Grass.

flexuosa, Muhl. Bogs, not rare.

"   "   var. pusilla, Gray.   "   "

JUNCACEÆ.

JUNCUS, Tourn. Rush.

effusus, L. Common Rush. Very common.

marginatus, Rostk. Moist sandy places.

"   "   var. *paucicapitatus, Engelm.

Princeton. *W. W. B. & J. F. C.*

tenuis, Willd.   Common.
pelocarpus, E. Meyer. Westminster. *Miss M. B. White.*
militaris. Bigel.   Bogs and streams. Uxbridge.
acuminatus, Michx.   Common.
Canadensis, J. Gay.   Common.
    "   "   " var. coarctatus, Engelm.
               Princeton. *W. W. B. & J. F. C.*
Luzula, DC.   Wood-Rush.
vernalis, DC.   Mt. Wachusett.
spadicea, DC., var. melanocarpa, Meyer. "   "
campestris, DC.   Woods, common.

TYPHACEÆ.

Typha, Tourn.   Cat-tail Flag.
latifolia, L.   Common Cat-tail.   Marshes, common.
Sparganium, Tourn.   Bur-reed.
eurycarpum, Engelm.   Borders of ponds, common.
simplex, Huds.   Southbridge. *L. E. Ammidown.*
    "   " var. androcladum, Engelm. *G. E. Stone.*
    "   "   " angustifolium,   "   " " "

ARACEÆ.

Arisæma, Martius.   Indian Turnip.
triphyllum, Torr. Jack-in-the-Pulpit.   Common.
Peltandra, Raf.   Arrow Arum.
undulata, Raf.   Shallow water. Sutton.
Calla, L.   Water Arum.
palustris, L.   Cold bogs, common.
Symplocarpus, Salisb.   Skunk Cabbage.
foetidus, Salisb.   Moist grounds, common.

Acorus, L.   Sweet Flag.
    Calamus, L.   Moist grounds, not rare.

LEMNACEÆ.

Lemna, L.   Duckweed.
    minor, L.   Stagnant waters, common.

ALISMACEÆ.

Alisma, L.   Water-Plantain.
    Plantago, L.   Shallow water, common.
Sagittaria, L.   Arrow-head.
    variabilis, Engelm.   In water or wet places.   Common.
      "   "   var. *obtusa.   *G. E. Stone.*
      "   "   " *angustifolia.   "   "   "
      "   "   " *diversifolia.   Princeton.
                        W. W. B. & J. F. C.
      "   "   " gracilis, Engelm.   *G. E. Stone.*
    graminea, Michx.   "   "   "

NAIADACEÆ.

Potamogeton, Tourn.   Pondweed. (By the late Rev. Thos.
          Morong.)
    natans, L.   Very common.
    "   " var. *prolixus, Koch.   "   "
    Oakesianus, Robbins.   Uxbridge.
    Pennsylvanicus, Cham.   Common in streams and ponds.
    Vaseyi, Robbins.   Lake Quinsigamond.
    Spirillus, Tuckerm.   Common.
    hybridus, Michx.   "
    fluitans, Roth.   Worcester.   *Miss E. W. Sargent.*

amplifolius, Tuckerm. Lake Quinsigamond.
heterophyllus, Schreb. "   "
    "   "   var. myriophyllus, Robbins.
                Lake Quinsigamond.
obtusifolius, Mertens & Koch. Beaver Brook, Worcester.
pauciflorus, Pursh. Lake Quinsigamond.
pusillus, L. Common.
    "   " var. tenuissimus, Koch. "
gemmiparus, Robbins. "
Tuckermani, Robbins. Shockalog Pond, Uxbridge.
Robbinsii, Oakes. Lake Quinsigamond.
NAIAS, L. Naiad.
flexilis, Rostk & Schmidt. Lake Quinsigamond.
                    T. Morong.
Indica, Cham., var. gracillima, A. Br. Lake Quinsigamond.
                    T. Morong.

ERIOCAULEÆ.

ERIOCAULON, L. Pipewort.
septangulare, With. Borders of ponds, not rare.

CYPERACEÆ.

CYPERUS, Tourn. Galingale.
diandrus, Torr. Low grounds.
filiculmis, Vahl. Dry soil.
dentatus, Torr. Margins of ponds, not rare.
strigosus, L. Swamps.
DULICHIUM, Pers. Dulichium.
spathaceum, Pers. Swamps.

ELEOCHARIS, R. Br.   Spike-Rush.
    ovata, R. Br.   Swamps.
    palustris, R. Br.   Ponds.
    intermedia, Schultes.
    tenuis, Schultes.   Margins of ponds.
    acicularis, R. Br.   Worcester. *Prof. Eaton.*
    pygmæa, Torr.   *G. E. Stone.*
FIMBRISTYLIS, Vahl.   Fimbristylis.
    autumnalis, Roem. & Schultes.   Low grounds.
    capillaris, Gray.   Dry soil.
SCIRPUS, Tourn.   Bulrush.
    subterminalis, Torr.   Slow streams.
    lacustris, L.   Great Bulrush.   Ponds and streams.
    sylvaticus, L.   Brooks. *Prof. Eaton.*
    atrovirens, Muhl.   Wet places, common.
    polyphyllus, Vahl.   Worcester. *Prof. Eaton.*
ERIOPHORUM, L.   Cotton-Grass.
    cyperinum, L.   Swamps, common.
    alpinum, L.   Eastern part of the county.
    vaginatum, L.   Worcester. *G. Coult.*
    Virginicum, L.   Low meadows, common.
    polystachyon, L.   "   "   "
      "   " var. *latifolium, Gray.   Bogs, not rare.
    gracile, Koch.   "   "   "
RHYNCHOSPORA, Vahl.   Beak-Rush.
    glomerata, Vahl.   Low grounds, not rare.
CAREX, Ruppius.   Sedge.
    folliculata, L.   Quite common.
    intumescens, Rudge.   Meadows.

lupulina, Muhl.   Swamps.
monile, Tuckerm.   Meadows.
bullata, Schkuhr.   "
lurida, Wahl.   "
Pseudo-Cyperus, L.   Fitchburg.   *Prof. Eaton.*
      "   "   " var. Americana, Hochst.   Millbury.
vestita, Willd.   Meadows.
filiformis, L.   Worcester.   *Prof. Eaton.*
      "   " var. latifolia, Bœckl.   "   "   "
trichocarpa, Muhl.   "   "   "
riparia, W. Curtis.   Swamps.
fusca, All.   Bogs. Southbridge.
stricta, Lam.   "
prasina, Wahl.   Meadows and bogs.   Not common.
crinita, Lam.   Bogs, common.
virescens, Muhl.   Copses.
      "   "   var. *costata, Dewey.   Princeton.
                                    W. W. B. & J. F. C.
triceps, Michx., var. hirsuta, Bailey.   Woods.
longirostris, Torr.   Princeton.   *W. W. B. & J. F. C.*
arctata, Boott.   Woods and copses.
debilis, Michx., var. Rudgei, Bailey.   Swamps.
gracillima, Schwein.   Woodlands, common.
grisea, Wahl.   Moist grounds.
flava, L.   Worcester.   *Prof. Eaton.*
pallescens, L.   Glades and meadows, common.
conoidea, Schkuhr.   Grassy places.
laxiflora, Lam.   Grassy places, common.
      "   "   var. *latifolia, Boott.   Princeton.
                                    W. W. B. & J. F. C.

digitalis, Willd.   Dry woods. *G. E. Stone.*

platyphylla, Carey.   Southbridge. *L. E. Ammidown.*

plantaginea, Lam.   "   " "   "

tetanica, Schkuhr.   Wet meadows.

varia, Muhl.   Dry woods.

Pennsylvanica, Lam.   Dry fields, common.

umbellata, Schkuhr.   Grassy knolls. Not common.

pubescens, Muhl.   Copses. *G. E. Stone.*

Willdenovii, Schkuhr.   Worcester. *Prof. Eaton.*

polytrichoides, Muhl.   Swamps.

stipata, Muhl.   Roadsides. Not rare.

teretiuscula, Gooden.   Worcester. *Prof. Eaton.*

vulpinoidea, Michx.   Meadows.

rosea, Schkuhr.   Rich woods.

"   "   var. *radiata, Dewey. Princeton.

                           W. W. B. & J. F. C.

sparganioides, Muhl.   Meadows.

*muricata, L.   Princeton. *W. W. B. & J. F. C.*

Muhlenbergii, Schkuhr.   Open sterile soil.

cephaloidea, Dewey.   Wet meadows.

cephalophora, Muhl.   Dry soil.

echinata, Murray, var. cephalantha, Bailey.   Rare.

"   "   var. *microstachys, Boeckl.

           Princeton. *W. W. B. & J. F. C.*

canescens, L.   Swamps.

"   " var. vulgaris, Bailey.   Common.

trisperma, Dewey.   Swamps.

Deweyana, Schwein.   Worcester. *Prof. Eaton.*

bromoides, Schkuhr.   Swampy woods.

tribuloides, Wahl., var. cristata, Bailey.   Worcester.

Prof. Eaton.

scoparia, Schkuhr.   Meadows.

straminea, Willd.   Dry fields.   *Prof. Eaton.*

"   "   var. mirabilis, Tuckerm.

Princeton.   *IV. W. B. & J. F. C.*

GRAMINEÆ.

PASPALUM, L.   Paspalum.

setaceum, Michx.   Pastures.

læve, Michx.   *G. E. Stone.*

PANICUM, L.   Panic-Grass.

*glabrum, Gaudin.   Pastures.

*sanguinale, L.   Common Crab-Grass.   Common.

capillare, L.   Old-witch Grass.   "

agrostoides, Muhl.   Sandy margins of ponds.

xanthophysum, Gray.   Princeton.

latifolium, L.   ·   Moist thickets.

depauperatum, Muhl.   Dry hills.

dichotomum, L.   Common.

*Crus-galli, L.   Barnyard-Grass.   "

*miliaceum, L.   East Worcester.

SETARIA, Beauv.   Bristly Foxtail Grass.

*glauca, Beauv.   Foxtail.   Not rare.

*viridis, Beauv.   Green Foxtail.   "   "

*Italica, Kunth.   Hungarian Grass.   Rarely spontaneous.

CENCHRUS, L.   Bur-Grass.

tribuloides, L.   Sandy soil.

LEERSIA, Swartz.   White Grass.

oryzoides, Swartz.   Rice Cut-grass.   Swamps.

ANDROPOGON, Royen.   Beard-Grass.
    furcatus, Muhl.   Dry soil. Not rare.
    scoparius, Michx.   Poor soil. Common.
    Virginicus, L.   Princeton. *Prof. Eaton.*
CHRYSOPOGON, Trin.   Broom-Corn.
    nutans, Benth. Wood-Grass.   Quite common.
PHALARIS, L.   Canary-Grass.
    *Canariensis, L. Canary-Grass.   Waste places.
    arundinacea, L. Reed Canary-Grass.   Wet grounds.
ANTHOXANTHUM, L.   Sweet Vernal-Grass.
    *odoratum, L.   Meadows, common.
STIPA, L.   Feather-Grass.
    Richardsonii, Link.   Mt. Wachusett. *Prof. Eaton.*
ORYZOPSIS, Michx.   Mountain Rice.
    melanocarpa, Muhl.   Princeton. *W. W. B. & J. F. C.*
    asperifolia, Michx.   Mt. Wachusett.
    Canadensis, Torr.   West Boylston. *Prof. Eaton.*
MUHLENBERGIA, Schreber.   Drop-seed Grass.
    glomerata, Trin.   Southbridge. *L. E. Ammidown.*
    Mexicana, Trin.   Princeton. *W. W. B. & J. F. C.*
    sylvatica, Torr. & Gray.   Copses, common.
    Willdenovii, Trin.   "   not rare.
    diffusa, Schreber. Drop-seed.   Dry hills.
BRACHYELYTRUM, Beauv.
    aristatum, Beauv.   Copses.
PHLEUM, L.   Cat's-tail Grass.
    *pratense, L. Timothy.   Common.
ALOPECURUS, L.   Foxtail Grass.
    *pratensis, L. Meadow Foxtail.   Common.

geniculatus, L., var. aristulatus, Torr.   In very wet places.

SPOROBOLUS, R. Br.   Drop-seed Grass.

vaginæflorus, Vasey.   Dry fields.

AGROSTIS, L.   Bent-Grass.

*alba, L.   White Bent-Grass.   Common.

* "   " var. vulgaris, Thurber.   Red Top.   "

perennans, Tuckerm.   Thin-Grass.   Damp places.

scabra, Willd.   Hair-Grass.   Dry places.

CINNA, L.   Wood Reed-Grass.

arundinacea, L.   Woods, not rare.

pendula, Trin.   Princeton.   W. W. B. & J. F. C.

CALAMAGROSTIS, Adans.   Reed Bent-Grass.

Canadensis, Beauv.   Blue-Joint Grass.   Meadows.

Nuttalliana, Steud.   Moist grounds.

ARRHENATHERUM, Beauv.   Oat-Grass.

*avenaceum, Beauv.   Meadows.

HOLCUS, L.   Meadow Soft-Grass.

*lanatus, L.   Velvet-Grass.   Moist meadows.

DESCHAMPSIA, Beauv.   Hair-Grass.

flexuosa, Trin.   Common Hair-Grass.   Mt. Wachusett.

AVENA, Tourn.   Oat.

*sativa, L.   Occasionally spontaneous.

striata, Michx.   G. E. Stone.

DANTHONIA, DC.   Wild Oat-Grass.

spicata, Beauv.   Pastures.

PHRAGMITES, Trin.   Reed.

communis, Trin.   Westborough.   A. N. Randlett.

ERAGROSTIS, Beauv.

*major, Host.   Waste places.

Purshii, Schrader.   *G. E. Stone.*
pectinacea, Gray.   Pastures.

DACTYLIS, L.   Orchard Grass.
*glomerata, L.   Common.

BRIZA, L.   Quaking Grass.
*media, L.   Pastures.

POA, L.   Meadow-Grass.
*annua, L.   Low Spear-Grass.   Common.
serotina, Ehrh.   False Red-top.   "
pratensis, L.   June Grass.   "

GLYCERIA, R. Br.   Manna-Grass.
Canadensis, Trin.   Rattlesnake-Grass.   Wet places.
obtusa, Trin.   "   "
nervata, Trin.   Fowl Meadow-Grass.   Moist meadows.
fluitans. R. Br.   Shallow water.
acutiflora, Torr.   Wet places.

FESTUCA, L.   Fescue-Grass.
ovina, L.   Sheep's Fescue.   *G. E. Stone.*
nutans, Willd.   Princeton.   *W. W. B. & J. F. C.*
*elatior, L.   Meadow Fescue.   Rich grassland.
"   " var. pratensis, Gray. Princeton.
                   W. W. B. & J. F. C.

BROMUS, L.   Brome-Grass.
*secalinus, L.   Cheat.   Waste grounds.
*racemosus, L.   Upright Chess.   "   "
ciliatus, L.   River banks, etc.
*asper, L.   Princeton.   *W. W. B. & J. F. C.*
*tectorum, L.   Waste grounds.

LOLIUM, L.   Darnel.
perenne, L.   Common Darnel.   Fields, etc.

AGROPYRUM, Gærtn.   Wheat-Grass.
repens, Beauv.   Quitch-Grass.   Common.
ELYMUS, L.   Wild Rye.
Virginicus, L.   Wet places.
Canadensis, L.   "   "
ASPRELLA, Willd.   Bottle-brush Grass.
Hystrix, Willd.   Moist woodlands.

SERIES II. CRYPTOGAMIA.

CLASS III. ACROGENÆ.

SUBCLASS I. PTERIDOPHYTA.

EQUISETACEÆ.

EQUISETUM, L.   Horsetail.
arvense, L.   Common Horsetail.   Gravelly soil.
sylvaticum, L.   Moist shaded places.
limosum, L.   In shallow water.
hyemale, L.   *G. E. Stone.*
scirpoides, Michx.   Southbridge.   *L. E. Ammidown.*

FILICES.

POLYPODIUM, L.   Polypody.
vulgare, L.   Rocks, common.
ADIANTUM, L.   Maidenhair.
pedatum, L.   Rich moist woods.
PTERIS, L.   Brake.
aquilina, L.   Common Brake.   Thickets, common.
ASPLENIUM, L.   Spleenwort.
Trichomanes, L.   Rare.

ebeneum, Ait.   Common.
thelypteroides, Michx.   "
Filix-fœmina, Bernh.   "
PHEGOPTERIS, Fée.   Beech Fern.
polypodioides, Fée.   Damp woods.
hexagonoptera, Fée.   "   "
Dryopteris, Fée.   "   "
ASPIDIUM, Swartz.   Shield Fern.
Thelypteris, Swartz.   Woodlands, common.
Noveboracense, Swartz.   "   "
spinulosum, Swartz, var. intermedium, D. C. Eaton.
                                                Common.
cristatum, Swartz.   Quite   "
   "   "   var. Clintonianum.   Worcester.
Goldianum, Hook.   Spencer. *Miss A. E. Tucker.*
marginale, Swartz.   Rich woods, common.
acrostichoides, Swartz. Christmas Fern.   "
CYSTOPTERIS, Bernh.   Bladder Fern.
fragilis, Bernh.   Worcester.
ONOCLEA, L.   Onoclea.
sensibilis, L. Sensitive Fern.   Very common.
   "   " var. *obtusilobata. Princeton.
                                  W. W. B. & J. F. C.
 ∙′   Struthiopteris, Hoffm. Ostrich Fern.   Worcester.
WOODSIA, R. Br.   Woodsia.
Ilvensis, R. Br.   Spencer. *Miss A. E. Tucker.*
obtusa, Torr.   *Miss Wheelock.*
DICKSONIA, L'Her.   Dicksonia.
pilosiuscula, Willd.   Moist shady places, common.

LYGODIUM, Swartz.   Climbing Fern.

palmatum, Swartz.   Uxbridge, Oxford.   Rare.

OSMUNDA, L.   Flowering Fern.

regalis, L.   Swamps, common.

Claytoniana, L.   Low grounds,   "

cinnamomea, L.   Cinnamon Fern.   "

OPHIOGLOSSACEÆ.

BOTRYCHIUM, Swartz.   Moonwort.

lanceolatum, Angstrœm.   Princeton. *W. W. B. & J. F. C.*

matricariæfolium, Braun.   "   "   "   "

ternatum, Swartz.   Pastures.

"   "   var. lunarioides.   "

"   "   " intermedium.   "

"   "   " dissectum.   "

Virginianum, Swartz.   Rich woods.

OPHIOGLOSSUM, L.   Adder's Tongue.

vulgatum, L.   Worcester. Spencer, *Miss A. E. Tucker.*

LYCOPODIACEÆ.

LYCOPODIUM, L.   Club-Moss.

Selago, L.   Mt. Watatic.

lucidulum, Michx. Staghorn Moss.   Common.

inundatum, L.   Sandy shores.

annotinum, L.   Leicester. *G. Coult.*

obscurum, L.   Moist woods, common.

clavatum, L.   Common Club-Moss.   "

complanatum, L. Ground Pine.   "

"   " var. *Chamæcyparissus. Leicester.

G. Coult.

SELAGINELLACEÆ.

SELAGINELLA, Beauv.·   Dwarf Club-Moss.
    rupestris, Spring.   Worcester.   *G. T. Rignel.*
    apus, Spring.   Southbridge.   *L. E. Ammidown.*
ISOETES, L.   Quillwort.
    lacustris, L.   *G. E. Stone.*
    echinospora, Durieu, var. Braunii, Engelm.   "   "   "
    riparia, Engelm.   "   "   "

NOTE.

THE names of the following species were either accidentally omitted or received too late to be inserted in the proper places indicated below :

PAGE 12.

HYPERICUM, Tourn.
    Canadense, L., var. *majus, Gray.   Princeton.
                      W. W. B. & J. F. C.
ABUTILON, Tourn.   Indian Mallow.
    *Avicennæ, Gærtn.   Velvet-Leaf.   Escaped.
               Southbridge.   *L. E. Ammidown.*
HIBISCUS, L.   Rose Mallow.
    *Trionum, L.   Bladder Ketmia.   Southbridge.
                   L. E. Ammidown.

PAGE 14.

VITIS, Tourn.
    æstivalis, Michx.   Summer Grape.   Common.

PAGE 17.

LATHYRUS, Tourn.
 palustris, L.   Berlin. *Miss J. M. Nichols.*

PAGE 21.

CALLITRICHE, L.
 heterophylla, Pursh. Mt. Wachusett. *W. W. B. & J. F. C.*

PAGE 34.

RHODODENDRON, L.
 viscosum, Torr., var. *glaucum, Gray.
         Princeton. *W. W. B. & J. F. C.*

PAGE 47.

QUERCUS, L.
 palustris, DuRoi. Pin Oak.   In the southern part of
 the county.

The hope expressed on page 33 concerning *Andromeda polifolia*, L. has been fulfilled since that page was printed. October 18, 1894, Mr. G. S. Newcomb of Westborough reported to me the finding of this shrub in a swamp close by that village, and a few days later I had the pleasure of visiting the locality and of finding it in abundance.

*** I shall be glad to receive any information as to the discovery of species hitherto unrecorded within our limits, or any facts concerning the distribution of our rarer species already recorded.

The preceding list contains 1,098 species and varieties, of which 55 are cryptogams.

Andromeda polifolia. *L.*
The Water **Andromeda.**

The maid Andromeda, divinely fair,
Forever lives in poesy: a group
Of stars in northern skies keeps bright her fame
This little flower each spring recalls her name.

APPENDIX.

The trees, shrubs and evergreen flowering plants growing without cultivation in Worcester County.

Coptis trifolia, Salisb.	Goldthread.
Liriodendron Tulipifera, L.	Tulip-Tree.
Berberis vulgaris, L.	Barberry.
Tilia Americana, L.	Basswood.
Xanthoxylum Americanum, Mill.	Northern Prickly Ash.
Ilex verticillata, Gray.	Black Alder.
laevigata, Gray.	Smooth Winterberry.
Nemopanthes fascicularis, Raf.	Mountain Holly.
Celastrus scandens, L.	Climbing Bitter-sweet.
Rhamnus cathartica, L.	Common Buckthorn.
Ceanothus Americanus, L.	New Jersey Tea.
Vitis Labrusca, L.	Northern Fox-Grape.
aestivalis, Michx.	Summer Grape.
cordifolia, Michx.	Frost Grape.
riparia, Michx.	River Grape.
Ampelopsis quinquefolia, Michx.	Virginian Creeper.
Acer Pennsylvanicum, L.	Striped Maple.
spicatum, Lam.	Mountain "
saccharinum, Wang.	Rock "

dasycarpum, Ehrb.	White Maple.
rubrum, L.	Red "
Rhus typhina, L.	Staghorn Sumach.
glabra, L.	Smooth "
copallina, L.	Dwarf "
venenata, DC.	Poison Dogwood.
Toxicodendron. L.	Poison Ivy.
" " var. radicans, L.	" "
Genista tinctoria, L.	Dyer's Greenweed.
Robinia Pseudacacia, L.	Common Locust.
viscosa, Vent.	Clammy "
Prunus pumila, L.	Dwarf Cherry.
Pennsylvanica, L. f.	Wild Red "
Virginiana, L.	Choke "
serotina, Ehrh.	Wild Black "
Spiræa salicifolia, L.	Common Meadow-Sweet.
tomentosa, L.	Hardhack.
Physocarpus opulifolius, Maxim.	Nine-bark.
Rubus odoratus, L.	Purple Flowering-Raspberry.
strigosus, Michx.	Wild Red "
occidentalis, L.	Thimbleberry.
villosus, Ait.	High Blackberry.
Canadensis, L.	Low "
hispidus, L.	Swamp "
Potentilla fruticosa, L.	Shrubby Cinquefoil.
Rosa blanda, Ait.	Wild Rose.
Carolina, L.	" "
lucida, Ehrh.	" "
humilis, Marsh.	" "

rubiginosa, L.	Sweetbrier.
Pyrus arbutifolia, L. f.	Choke-berry.
" " " var. melanocarpa, Hook.	"
Americana, DC.	American Mountain-Ash.
aucuparia, Gærtn.	European "
Malus, L.	Wild Apple.
communis, L.	" Pear.
Cratægus coccinea, L.	White Thorn.
" " var. mollis, Torr. & Gray.	" "
punctata, Jacq.	" "
Amelanchier Canadensis, Torr. & Gray.	Shad-bush.
Hydrangea arborescens, L.	Wild Hydrangea.
Ribes Cynosbati, L.	Gooseberry.
rotundifolium, Michx.	"
oxyacanthoides, L.	"
prostratum, L'Her.	Fetid Currant.
floridum, L'Her.	Wild Black "
rubrum, L., var. subglandulosum, Maxim.	Red "
Hamamelis Virginiana, L.	Witch-Hazel.
Decodon verticillatus, Ell.	Swamp Loosestrife.
Aralia hispida, Vent.	Bristly Sarsaparilla.
Cornus florida, L.	Flowering Dogwood.
circinata, L'Her.	Round-leaved Cornel.
sericea, L.	Silky "
stolonifera, Michx.	Red-osier Dogwood.
paniculata, L'Her.	Panicled Cornel.
alternifolia, L. f.	Alternate-leaved "
Nyssa sylvatica, Marsh.	Tupelo.

Sambucus Canadensis, L.	Common Elder.
racemosa, L.	Red-berried "
Viburnum lantanoides, Michx.	Hobble-bush.
Opulus, L.	Cranberry-tree.
acerifolium, L.	Arrow-wood.
dentatum, L.	"
cassinoides, L.	Withe-rod.
Lentago, L.	Sweet Viburnum.
Linnæa borealis, Gronov.	Twin-flower.
Symphoricarpus racemosus, Michx.	Snowberry.
Lonicera ciliata, Muhl.	Fly-Honeysuckle.
cærulea, L.	Mountain " "
hirsuta, Eaton.	Hairy "
glauca, Hill.	"
Diervilla trifida, Mœnch.	Bush Honeysuckle.
Cephalanthus occidentalis, L.	Button-bush.
Mitchella repens, L.	Partridge-berry.
Gaylussacia dumosa, Torr. & Gray.	Dwarf Huckleberry.
frondosa, Torr. & Gray.	Dangleberry.
resinosa, Torr. & Gray.	Huckleberry.
Vaccinium Pennsylvanicum, Lam.	Dwarf Blueberry.
Canadense, Kalm.	Canada "
vacillans, Solander.	Low "
corymbosum, L. & vars.	High "
Oxycoccus, L.	Small Cranberry.
macrocarpon, Ait.	Large "
Chiogenes serpyllifolia, Salisb.	Creeping Snowberry.
Arctostaphylos Uva-ursi, Spreng.	Bearberry.
Epigæa repens, L.	Trailing Arbutus.
Gaultheria procumbens, L.	Checkerberry.

Rhododendron nudiflorum, Torrey.
The Swamp Pink.

" What splendid masses of pink, with a few glaucous green leaves sprinkled here and there—just enough for contrast."

—H. D. Thoreau

Andromeda polifolia, L.	Water Andromeda.
ligustrina, Muhl.	Andromeda.
Leucothoë racemosa, Gray.	Leucothoë.
Cassandra calyculata, Don.	Leather-Leaf.
Kalmia latifolia, L.	Mountain Laurel.
angustifolia, L.	Sheep "
glauca, Ait.	Pale "
Rhododendron viscosum, Torr.	White Azalea.
" " var. glaucum, Gray.	
	White Azalea.
nudiflorum, Torr.	Swamp Pink.
Rhodora, Don.	Rhodora.
maximum, L.	Rhododendron.
Ledum latifolium, Ait.	Labrador Tea.
Clethra alnifolia, L.	Sweet Pepperbush.
Chimaphila umbellata, Nutt.	Prince's Pine.
maculata, Pursh.	Spotted Wintergreen.
Moneses grandiflora, Salisb.	One-flowered Pyrola.
Pyrola secunda, L.	Wintergreen.
chlorantha, Swartz.	"
elliptica. Nutt.	Shin-leaf.
rotundifolia, L.	Wintergreen.
Fraxinus Americana, L.	White Ash.
sambucifolia, Lam.	Black Ash.
Sassafras officinale, Nees.	Sassafras.
Lindera Benzoin, Blume.	Spice-bush.
Dirca palustris, L.	Leatherwood.
Daphne Mezereum, L.	Mezereum.
Ulmus fulva, Michx.	Slippery Elm.
Americana, L.	American "

Celtis occidentalis, L.   Hackberry.

Mr. G. A. Cheney assures me that he has found it growing in Sturbridge and elsewhere in the county.

Morus alba, L.	White Mulberry.
Platanus occidentalis, L.	Buttonwood.
Juglans cinerea, L.	Butternut.
Carya alba, Nutt.	Shag-bark Hickory.
porcina, Nutt.	Pig-nut "
amara, Nutt.	Bitter-nut "
Myrica Gale, L.	Sweet Gale.
cerifera, L.	Bayberry.
asplenifolia, Endl.	Sweet Fern.
Betula lenta, L.	Black Birch.
lutea, Michx. f.	Yellow "
populifolia, Ait.	Gray "
papyrifera, Marshall.	Paper "
nigra, L.	Red "
Alnus incana, Willd.	Speckled Alder.
serrulata, Willd.	Smooth "
Corylus Americana, Walt.	Wild Hazel-nut.
rostrata, Ait.	Beaked "
Ostrya Virginica, Willd.	Hop-Hornbeam.
Carpinus Caroliniana, Walt.	Hornbeam.
Quercus alba, L.	White Oak.
macrocarpa, Michx.	Bur "

I am indebted to Mrs. H. G. Waite for the addition of this species.

bicolor, Willd.	Swamp White Oak.
Prinus, L.	Chestnut "
" " var. monticola, Michx.	Rock " "

prinoides, Willd.	Dwarf Chestnut Oak.
rubra, L.	Red "
coccinea, Wang.	Scarlet "
" " var. tinctoria, Gray.	Black "
palustris, DuRoi.	Pin "
Miss A. H. Tucker in *Trees of Worcester.*	
ilicifolia, Wang.	Scrub "
Castanea sativa, Mill., var. Americana, Michx.	Chestnut.
Fagus ferruginea, Ait.	Beech.
Salix nigra, Marsh.	Black Willow.
lucida, Muhl.	Shining "
fragilis, L.	Crack "
alba, L.	White "
" " var. vitellina, Koch.	" "
rostrata, Richardson.	Beaked "
discolor, Muhl.	Glaucous "
humilis, Marsh.	Prairie "
tristis, Ait.	Dwarf Gray "
sericea, Marsh.	Silky "
cordata, Muhl.	Heart-leaved "
myrtilloides, L.	"
Populus tremuloides, Michx.	American Aspen.
grandidentata, Michx.	Large-toothed "
balsamifera, L., var. candicans, Gray.	Balm of Gilead.
monilifera, Ait.	Cottonwood.
Pinus Strobus, L.	White Pine.
rigida, Mill.	Pitch "
resinosa, Ait.	Red "
Picea nigra, Link.	Black Spruce.

Tsuga Canadensis, Carr.	Hemlock.
Abies balsamea, Mill.	Balsam Fir.
Larix Americana, Michx.	Hackmatack.
Chamæcyparis sphæroidea, Spach.	White Cedar.
Juniperus communis, L.	Common Juniper.
Sabina, L., var. procumbens, Pursh.	. "
Virginiana, L.	Red Cedar.
Taxus Canadensis, Willd.	Ground Hemlock.
Smilax rotundifolia, L.	Common Greenbrier.

Notes.